Networking Wireless Sensors

Wireless sensor networks promise an unprecedented fine-grained interface between the virtual and physical worlds. They are one of the most rapidly developing new information technologies, with applications in a wide range of fields including industrial process control, security and surveillance, environmental sensing, and structural health monitoring. This book is motivated by the urgent need to provide a comprehensive and organized survey of the field. It shows how the core challenges of energy efficiency, robustness, and autonomy are addressed in these systems by networking techniques across multiple layers. The topics covered include network deployment, localization, time synchronization, wireless radio characteristics, medium-access, topology control, routing, data-centric techniques, and transport protocols.

Ideal for researchers and designers seeking to create new algorithms and protocols and engineers implementing integrated solutions, it also contains many exercises and can be used by graduate students taking courses in networks.

BHASKAR KRISHNAMACHARI is an assistant professor in the Department of Electrical Engineering Systems at the University of Southern California.

Networking Wireless Sensors

Bhaskar Krishnamachari

Deployment & configuration – Localization

Synchronization – Wireless characteristics

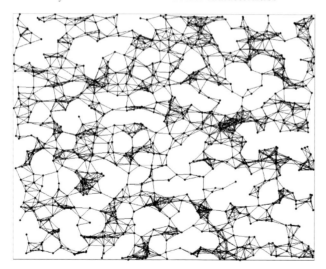

Sleep-oriented MAC – Efficient routing

Data-centric concepts – Congestion control

CAMBRIDGE
UNIVERSITY PRESS

CAMBRIDGE UNIVERSITY PRESS
Cambridge, New York, Melbourne, Madrid, Cape Town, Singapore, São Paulo

CAMBRIDGE UNIVERSITY PRESS
The Edinburgh Building, Cambridge CB2 2RU, UK

Published in the United States of America by Cambridge University Press, New York

www.cambridge.org
Information on this title: www.cambridge.org/9780521838474

First published 2005

Printed in the United Kingdom at the University Press, Cambridge

Typeface 10/13pt. Times and Formata   *System* LaTeX 2ε

A catalog record for this publication is available from the British Library

ISBN-13 978-0-521-83847-4 hardback
ISBN-10 0-521-83847-9 hardback

To Shriram & Zhen,
Amma & Appa

Contents

Preface

Every piece of honest writing contains this tacit message: "I wrote this because it is important; I want you to read it; I'll stand behind it."

Matthew Grieder, as quoted by J.R. Trimble, in *Writing with Style*

With its origins in the early nineties, the subject of wireless sensor networks has seen an explosive growth in interest in both academia and industry. In just the past five years several hundred papers have been written on the subject. I have written this book because I believe there is an urgent need to make this vast literature more readily accessible to students, researchers, and design engineers.

The book aims to provide the reader with a comprehensive, organized survey of the many protocols and fundamental design concepts developed for wireless sensor networks in recent years. The topics covered are wide-ranging: deployment, localization, synchronization, wireless link characteristics, medium-access, sleep scheduling and topology control, routing, data-centric concepts, and congestion control.

This book has its origins in notes, lectures, and discussions from a graduate course on wireless sensor networks that I've taught thrice at the University of Southern California in the past two years. This text will be of interest to senior undergraduate and graduate students in electrical engineering, computer science, and related engineering disciplines, as well as researchers and practitioners in academia and industry.

To keep the book focused coherently on networking issues, I have had to limit in-depth treatment of some topics. These include target tracking, collaborative signal processing, distributed computation, programming and middleware, and security protocols. However, these topics are all addressed briefly in the final chapter, along with pointers to key relevant papers.

I am certain there is much room for improvement in this work. I would be delighted to receive suggestions from readers via e-mail to bkrishna@usc.edu.

Acknowledgements: First and foremost, I would like to thank Zhen, my wife, for her support through this whole writing process, and my little son Shriram, for making sure it wasn't all work and no play. My sincere thanks to the students in my wireless sensor networks classes at USC for many in-depth discussions on the subject; to my own graduate students from the Autonomous Networks Research Group for their considerable assistance; to the many faculty and researchers at USC and beyond who have offered useful advice and from whom I have learned so much; and to my editors at Cambridge University Press for all their patience and help.

Bhaskar Krishnamachari

Introduction

1.1 Wireless sensor networks: the vision

Recent technological advances allow us to envision a future where large numbers of low-power, inexpensive sensor devices are densely embedded in the physical environment, operating together in a wireless network. The envisioned applications of these wireless sensor networks range widely: ecological habitat monitoring, structure health monitoring, environmental contaminant detection, industrial process control, and military target tracking, among others.

A US National Research Council report titled *Embedded Everywhere* notes that the use of such networks throughout society "could well dwarf previous milestones in the information revolution" [47]. Wireless sensor networks provide bridges between the virtual world of information technology and the real physical world. They represent a fundamental paradigm shift from traditional inter-human personal communications to autonomous inter-device communications. They promise unprecedented new abilities to observe and understand large-scale, real-world phenomena at a fine spatio-temporal resolution. As a result, wireless sensor networks also have the potential to engender new breakthrough scientific advances.

While the notion of networking distributed sensors and their use in military and industrial applications dates back at least to the 1970s, the early systems were primarily wired and small in scale. It was only in the 1990s – when wireless technologies and low-power VLSI design became feasible – that researchers began envisioning and investigating large-scale embedded wireless sensor networks for dense sensing applications.

Figure 1.1 A Berkeley mote (MICAz MPR2400 series)

Perhaps one of the earliest research efforts in this direction was the low-power wireless integrated microsensors (LWIM) project at UCLA funded by DARPA [98]. The LWIM project focused on developing devices with low-power electronics in order to enable large, dense wireless sensor networks. This project was succeeded by the Wireless Integrated Networked Sensors (WINS) project a few years later, in which researchers at UCLA collaborated with Rockwell Science Center to develop some of the first wireless sensor devices. Other early projects in this area, starting around 1999–2000, were also primarily in academia, at several places including MIT, Berkeley, and USC.

Researchers at Berkeley developed embedded wireless sensor networking devices called motes, which were made publicly available commercially, along with TinyOS, an associated embedded operating system that facilitates the use of these devices [81]. Figure 1.1 shows an image of a Berkeley mote device. The availability of these devices as an easily programmable, fully functional, relatively inexpensive platform for experimentation, and real deployment has played a significant role in the ongoing wireless sensor networks revolution.

1.2 Networked wireless sensor devices

As shown in Figure 1.2, there are several key components that make up a typical wireless sensor network (WSN) device:

1. **Low-power embedded processor:** The computational tasks on a WSN device include the processing of both locally sensed information as well as information communicated by other sensors. At present, primarily due to economic

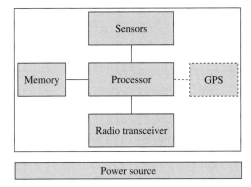

Figure 1.2 Schematic of a basic wireless sensor network device

constraints, the embedded processors are often significantly constrained in terms of computational power (e.g., many of the devices used currently in research and development have only an eight-bit 16-MHz processor). Due to the constraints of such processors, devices typically run specialized component-based embedded operating systems, such as TinyOS. However, it should be kept in mind that a sensor network may be heterogeneous and include at least some nodes with significantly greater computational power. Moreover, given Moore's law, future WSN devices may possess extremely powerful embedded processors. They will also incorporate advanced low-power design techniques, such as efficient sleep modes and dynamic voltage scaling to provide significant energy savings.

2. **Memory/storage:** Storage in the form of random access and read-only memory includes both program memory (from which instructions are executed by the processor), and data memory (for storing raw and processed sensor measurements and other local information). The quantities of memory and storage on board a WSN device are often limited primarily by economic considerations, and are also likely to improve over time.

3. **Radio transceiver:** WSN devices include a low-rate, short-range wireless radio (10–100 kbps, <100 m). While currently quite limited in capability too, these radios are likely to improve in sophistication over time – including improvements in cost, spectral efficiency, tunability, and immunity to noise, fading, and interference. Radio communication is often the most power-intensive operation in a WSN device, and hence the radio must incorporate energy-efficient sleep and wake-up modes.

4. **Sensors:** Due to bandwidth and power constraints, WSN devices primarily support only low-data-rate sensing. Many applications call for multi-modal sensing, so each device may have several sensors on board. The specific

sensors used are highly dependent on the application; for example, they may include temperature sensors, light sensors, humidity sensors, pressure sensors, accelerometers, magnetometers, chemical sensors, acoustic sensors, or even low-resolution imagers.

5. **Geopositioning system:** In many WSN applications, it is important for all sensor measurements to be location stamped. The simplest way to obtain positioning is to pre-configure sensor locations at deployment, but this may only be feasible in limited deployments. Particularly for outdoor operations, when the network is deployed in an *ad hoc* manner, such information is most easily obtained via satellite-based GPS. However, even in such applications, only a fraction of the nodes may be equipped with GPS capability, due to environmental and economic constraints. In this case, other nodes must obtain their locations indirectly through network localization algorithms.

6. **Power source:** For flexible deployment the WSN device is likely to be battery powered (e.g. using LiMH AA batteries). While some of the nodes may be wired to a continuous power source in some applications, and energy harvesting techniques may provide a degree of energy renewal in some cases, the finite battery energy is likely to be the most critical resource bottleneck in most WSN applications.

Depending on the application, WSN devices can be networked together in a number of ways. In basic data-gathering applications, for instance, there is a node referred to as the *sink* to which all data from *source* sensor nodes are directed. The simplest logical topology for communication of gathered data is a single-hop star topology, where all nodes send their data directly to the sink. In networks with lower transmit power settings or where nodes are deployed over a large area, a multi-hop tree structure may be used for data-gathering. In this case, some nodes may act both as sources themselves, as well as routers for other sources.

One interesting characteristic of wireless sensor networks is that they often allow for the possibility of intelligent in-network processing. Intermediate nodes along the path do not act merely as packet forwarders, but may also examine and process the content of the packets going through them. This is often done for the purpose of data compression or for signal processing to improve the quality of the collected information.

1.3 Applications of wireless sensor networks

The several envisioned applications of WSN are still very much under active research and development, in both academia and industry. We describe a few applications from different domains briefly to give a sense of the wide-ranging scope of this field:

1. **Ecological habitat monitoring:** Scientific studies of ecological habitats (animals, plants, micro-organisms) are traditionally conducted through hands-on field activities by the investigators. One serious concern in these studies is what is sometimes referred to as the "observer effect" – the very presence and potentially intrusive activities of the field investigators may affect the behavior of the organisms in the monitored habitat and thus bias the observed results. Unattended wireless sensor networks promise a cleaner, remote-observer approach to habitat monitoring. Further, sensor networks, due to their potentially large scale and high spatio-temporal density, can provide experimental data of an unprecedented richness.

   One of the earliest experimental deployments of wireless sensor networks was for habitat monitoring, on Great Duck Island, Maine [130]. A team of researchers from the Intel Research Lab at Berkeley, University of California at Berkeley, and the College of the Atlantic in Bar Harbor deployed wireless sensor nodes in and around burrows of Leach's storm petrel, a bird which forms a large colony on that island during the breeding season. The sensor-network-transmitted data were made available over the web, via a base station on the island connected to a satellite communication link.

2. **Military surveillance and target tracking:** As with many other information technologies, wireless sensor networks originated primarily in military-related research. Unattended sensor networks are envisioned as the key ingredient in moving towards network-centric warfare systems. They can be rapidly deployed for surveillance and used to provide battlefield intelligence regarding the location, numbers, movement, and identity of troops and vehicles, and for detection of chemical, biological, and nuclear weapons.

   Much of the impetus for the fast-growing research and development of wireless sensor networks has been provided though several programs funded by the US Defense Advanced Research Projects Agency (DARPA), most notably through a program known as Sensor Information Technology (SensIT) [188] from 1999 to 2002. Indeed, many of the leading US researchers and entrepreneurs in the area of wireless sensor networks today have been and are being funded by these DARPA programs.

3. **Structural and seismic monitoring:** Another class of applications for sensor networks pertains to monitoring the condition of civil structures [231]. The structures could be buildings, bridges, and roads; even aircraft. At present the health of such structures is monitored primarily through manual and visual

inspections or occasionally through expensive and time-consuming technologies, such as X-rays and ultrasound. Unattended networked sensing techniques
can automate the process, providing rich and timely information about incip-
ient cracks or about other structural damage. Researchers envision deploying
these sensors densely on the structure – either literally embedded into the
building material such as concrete, or on the surface. Such sensor networks
have potential for monitoring the long-term wear of structures as well as
their condition after destructive events, such as earthquakes or explosions.
A particularly compelling futuristic vision for the use of sensor networks
involves the development of controllable structures, which contain actuators
that react to real-time sensor information to perform "echo-cancellation" on
seismic waves so that the structure is unaffected by any external disturbance.

4. **Industrial and commercial networked sensing:** In industrial manufacturing
facilities, sensors and actuators are used for process monitoring and control.
For example, in a multi-stage chemical processing plant there may be sensors
placed at different points in the process in order to monitor the temperature,
chemical concentration, pressure, etc. The information from such real-time
monitoring may be used to vary process controls, such as adjusting the amount
of a particular ingredient or changing the heat settings. The key advantage
of creating wireless networks of sensors in these environments is that they
can significantly improve both the cost and the flexibility associated with
installing, maintaining, and upgrading wired systems [131]. As an indication
of the commercial promise of wireless embedded networks, it should be noted
that there are already several companies developing and marketing these
products, and there is a clear ongoing drive to develop related technology
standards, such as the IEEE 802.15.4 standard [94], and collaborative industry
efforts such as the Zigbee Alliance [244].

1.4 Key design challenges

Wireless sensor networks are interesting from an engineering perspective,
because they present a number of serious challenges that cannot be adequately
addressed by existing technologies:

1. **Extended lifetime:** As mentioned above, WSN nodes will generally be
severely energy constrained due to the limitations of batteries. A typical alka-
line battery, for example, provides about 50 watt-hours of energy; this may
translate to less than a month of continuous operation for each node in full
active mode. Given the expense and potential infeasibility of monitoring and

replacing batteries for a large network, much longer lifetimes are desired. In practice, it will be necessary in many applications to provide guarantees that a network of unattended wireless sensors can remain operational without any replacements for several years. Hardware improvements in battery design and energy harvesting techniques will offer only partial solutions. This is the reason that most protocol designs in wireless sensor networks are designed explicitly with energy efficiency as the primary goal. Naturally, this goal must be balanced against a number of other concerns.

2. **Responsiveness:** A simple solution to extending network lifetime is to operate the nodes in a duty-cycled manner with periodic switching between sleep and wake-up modes. While synchronization of such sleep schedules is challenging in itself, a larger concern is that arbitrarily long sleep periods can reduce the responsiveness and effectiveness of the sensors. In applications where it is critical that certain events in the environment be detected and reported rapidly, the latency induced by sleep schedules must be kept within strict bounds, even in the presence of network congestion.

3. **Robustness:** The vision of wireless sensor networks is to provide large-scale, yet fine-grained coverage. This motivates the use of large numbers of inexpensive devices. However, inexpensive devices can often be unreliable and prone to failures. Rates of device failure will also be high whenever the sensor devices are deployed in harsh or hostile environments. Protocol designs must therefore have built-in mechanisms to provide robustness. It is important to ensure that the global performance of the system is not sensitive to individual device failures. Further, it is often desirable that the performance of the system degrade as gracefully as possible with respect to component failures.

4. **Synergy:** Moore's law-type advances in technology have ensured that device capabilities in terms of processing power, memory, storage, radio transceiver performance, and even accuracy of sensing improve rapidly (given a fixed cost). However, if economic considerations dictate that the cost per node be reduced drastically from hundreds of dollars to less than a few cents, it is possible that the capabilities of individual nodes will remain constrained to some extent. The challenge is therefore to design synergistic protocols, which ensure that the system as a whole is more capable than the sum of the capabilities of its individual components. The protocols must provide an efficient collaborative use of storage, computation, and communication resources.

5. **Scalability:** For many envisioned applications, the combination of fine-granularity sensing and large coverage area implies that wireless sensor

networks have the potential to be extremely large scale (tens of thousands, perhaps even millions of nodes in the long term). Protocols will have to be inherently distributed, involving localized communication, and sensor networks must utilize hierarchical architectures in order to provide such scalability. However, visions of large numbers of nodes will remain unrealized in practice until some fundamental problems, such as failure handling and *in-situ* reprogramming, are addressed even in small settings involving tens to hundreds of nodes. There are also some fundamental limits on the throughput and capacity that impact the scalability of network performance.

6. **Heterogeneity:** There will be a heterogeneity of device capabilities (with respect to computation, communication, and sensing) in realistic settings. This heterogeneity can have a number of important design consequences. For instance, the presence of a small number of devices of higher computational capability along with a large number of low-capability devices can dictate a two-tier, cluster-based network architecture, and the presence of multiple sensing modalities requires pertinent sensor fusion techniques. A key challenge is often to determine the right combination of heterogeneous device capabilities for a given application.

7. **Self-configuration:** Because of their scale and the nature of their applications, wireless sensor networks are inherently *unattended* distributed systems. Autonomous operation of the network is therefore a key design challenge. From the very start, nodes in a wireless sensor network have to be able to configure their own network topology; localize, synchronize, and calibrate themselves; coordinate inter-node communication; and determine other important operating parameters.

8. **Self-optimization and adaptation:** Traditionally, most engineering systems are optimized *a priori* to operate efficiently in the face of expected or well-modeled operating conditions. In wireless sensor networks, there may often be significant uncertainty about operating conditions prior to deployment. Under such conditions, it is important that there be in-built mechanisms to autonomously learn from sensor and network measurements collected over time and to use this learning to continually improve performance. Also, besides being uncertain *a priori*, the environment in which the sensor network operates can change drastically over time. WSN protocols should also be able to adapt to such environmental dynamics in an online manner.

9. **Systematic design:** As we shall see, wireless sensor networks can often be highly application specific. There is a challenging tradeoff between (a) *ad hoc*, narrowly applicable approaches that exploit application-specific characteristics to offer performance gains and (b) more flexible, easy-to-generalize

design methodologies that sacrifice some performance. While performance optimization is very important, given the severe resource constraints in wireless sensor networks, systematic design methodologies, allowing for reuse, modularity, and run-time adaptation, are necessitated by practical considerations.

10. **Privacy and security:** The large scale, prevalence, and sensitivity of the information collected by wireless sensor networks (as well as their potential deployment in hostile locations) give rise to the final key challenge of ensuring both privacy and security.

1.5 Organization

This book is organized in a bottom–up manner. Chapter 2 addresses tools, techniques, and metrics pertinent to network deployment. Chapter 3 and Chapter 4 present techniques for spatial localization and temporal synchronization respectively. Chapter 5 addresses a number of issues pertaining to wireless characteristics, including models for link quality, interference, and radio energy. Algorithms for medium-access and radio sleep scheduling for energy conservation are described in Chapter 6. Topology control techniques based on sleep–active transitions are described in Chapter 7. Mechanisms for energy-efficient and robust routing are discussed in Chapter 8, while Chapter 9 presents concepts and techniques for data-centric routing and querying in wireless sensor networks. Chapter 10 covers issues pertinent to congestion control and transport-layer quality of service. Finally, we present concluding comments in Chapter 11, along with a brief survey of some important further topics.

2

Network deployment

2.1 Overview

The problem of *deployment* of a wireless sensor network could be formulated generically as follows: given a particular application context, an operational region, and a set of wireless sensor devices, how and where should these nodes be placed?

The network must be deployed keeping in mind two main objectives: coverage and connectivity. *Coverage* pertains to the application-specific quality of information obtained from the environment by the networked sensor devices. *Connectivity* pertains to the network topology over which information routing can take place. Other issues, such as equipment costs, energy limitations, and the need for robustness, should also be taken into account.

A number of basic questions must be considered when deploying a wireless sensor network:

1. **Structured versus randomized deployment:** Does the network involve (a) structured placement, either by hand or via autonomous robotic nodes, or (b) randomly scattered deployment?
2. **Over-deployment versus incremental deployment:** For robustness against node failures and energy depletion, should the network be deployed *a priori* with redundant nodes, or can nodes be added or replaced incrementally when the need arises? In the former case, sleep scheduling is desirable to extend network lifetime, a topic we will treat in Chapter 7.
3. **Network topology:** Is the network topology going to be a simple star topology, or a grid, or an arbitrary multi-hop mesh, or a two-level cluster hierarchy? What kind of robust connectivity guarantees are desired?

4. **Homogeneous versus heterogeneous deployment:** Are all sensor nodes of the same type or is there a mix of high- and low-capability devices? In case of heterogeneous deployments, there may be multiple gateway/sink devices (nodes to which sensor nodes report their data and through which an external user can access the sensor network).

5. **Coverage metrics:** What is the kind of sensor information desired from the environment and how is the coverage measured? This could be on the basis of detection and false alarm probabilities or whether every event can be sensed by K distinct nodes, etc.

We shall address these questions, beginning with the first.

2.2 Structured versus randomized deployment

The randomized deployment approach is appealing for futuristic applications of a large scale, where nodes are dropped from aircraft or mixed into concrete before being embedded in a smart structure. However, many small–medium-scale WSNs are likely to be deployed in a structured manner via careful hand placement of network nodes. In both cases, the cost and availability of equipment will often be a significant constraint.

We can illustrate these issues by considering in detail one possible methodology for structured placement:

1. Place sink/gateway device at a location that provides the desired wired network and power connectivity.
2. Place sensor nodes in a prioritized manner at locations of the operational area where sensor measurements are needed.
3. If necessary, add additional nodes to provide requisite network connectivity.

Step 2 can be challenging if it is not clear exactly where sensor measurements are needed, in which case a uniform or grid-like placement could be a suitable choice. Adding nodes for ensuring sufficient wireless network connectivity can also be a non-trivial challenge, particularly when there are location constraints in a given environment that dictate where nodes can or cannot be placed. If the number of available nodes is small with respect to the size of the operational area and required coverage, a delicate balance has to be struck between how many nodes can be allocated for sensor measurements and how many nodes are needed for routing connectivity.

Randomized sensor deployment can be even more challenging in some respects, since there is no way to configure *a priori* the exact location of each device. Additional post-deployment self-configuration mechanisms are therefore required to obtain the desired coverage and connectivity. In case of a uniform random deployment, the only parameters that can be controlled *a priori* are the numbers of nodes and some related settings on these nodes, such as their transmission range. We shall discuss some results from Random Graph Theory in Section 2.4 that provide useful insights into the settings of these parameters.

Regardless of whether the deployment is randomized or structured, the connectivity properties of the network topology can be further adjusted after deployment by varying transmit powers. We will discuss variable power-based topology control techniques in Section 2.5.

2.3 Network topology

The communication network can be configured into several different topologies, as seen in Figure 2.1. We describe these topologies below.

2.3.1 Single-hop star

The simplest WSN topology is the single-hop star shown in Figure 2.1(a). Every node in this topology communicates its measurements directly to the gateway. Wherever feasible, this approach can significantly simplify design, as the networking concerns are reduced to a minimum. However, the limitation of this topology is its poor scalability and robustness properties. For instance, in larger areas, nodes that are distant from the gateway will have poor-quality wireless links.

2.3.2 Multi-hop mesh and grid

For larger areas and networks, multi-hop routing is necessary. Depending on how they are placed, the nodes could form an arbitrary mesh graph as in Figure 2.1(b) or they could form a more structured communication graph such as the 2D grid structure shown in Figure 2.1(c).

2.3.3 Two-tier hierarchical cluster

Perhaps the most compelling architecture for WSN is a deployment architecture where multiple nodes within each local region report to different cluster-heads [76]. There are a number of ways in which such a hierarchical architecture

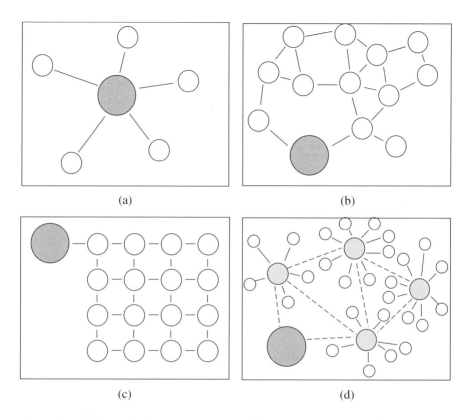

Figure 2.1 Different deployment topologies: (a) a star-connected single-hop topology, (b) flat multi-hop mesh, (c) structured grid, and (d) two-tier hierarchical cluster topology

may be implemented. This approach becomes particularly attractive in heterogeneous settings when the cluster-head nodes are more powerful in terms of computation/communication [90, 114]. The advantage of the hierarchical cluster-based approach is that it naturally decomposes a large network into separate zones within which data processing and aggregation can be performed locally. Within each cluster there could be either single-hop or multi-hop communication. Once data reach a cluster-head they would then be routed through the second-tier network formed by cluster-heads to another cluster-head or a gateway. The second-tier network may utilize a higher bandwidth radio or it could even be a wired network if the second-tier nodes can all be connected to the wired infrastructure. Having a wired network for the second tier is relatively easy in building-like environments, but not for random deployments in remote locations. In random deployments there may be no designated cluster-heads; these may have to be determined by some process of self-election.

2.4 Connectivity in geometric random graphs

The connectivity (and coverage) properties of random deployments can be best analyzed using Random Graph Theory. There are several models of random graphs that have been studied in the literature.

A *random graph model* is essentially a systematic description of some random experiment that can be used to generate graph instances. These models usually contain a tuning parameter that varies the average density of the constructed random graph. The Bernoulli random graphs $G(n, p)$, studied in traditional Random Graph Theory [11], are formed by taking n vertices and placing random edges between each pair of vertices independently with probability p.

A random graph model that more closely represents wireless multi-hop networks is the geometric random graph $G(n, R)$. In a $G(n, R)$ geometric random graph, n nodes are placed at random with uniform distribution in a square area of unit size (more generally, a d-dimensional cube). There is an edge (u, v) between any pair of nodes u and v, if the Euclidean distance between them is less than R.

Figure 2.2 illustrates $G(n, R)$ for $n = 40$ at two different R values. When R is small, each node can connect only to other nodes that are close by, and the resulting graph is sparse; on the other hand, a large R allows longer links and results in a dense connectivity.

Compared with Bernoulli random graphs, $G(n, R)$ geometric random graphs need different analytical techniques. This is because geometric random graphs do not show independence between edges. For instance, the probability that edge (u, v) exists is not independent of the probability that edge (u, w) and edge (v, w) exist.

2.4.1 Connectivity in *G(n, R)*

Figure 2.3 shows how the probability of network connectivity varies as the radius parameter R of a geometric random graph is varied. Depending on the number of nodes n, there exist different critical radii beyond which the graph is connected with high probability. These transitions become sharper (shifting to lower radii) as the number of nodes increases.

Figure 2.4 shows the probability that the network is connected with respect to the total number of nodes for different values of fixed transmission range in a fixed area for all nodes. It can be observed that, depending on the transmission

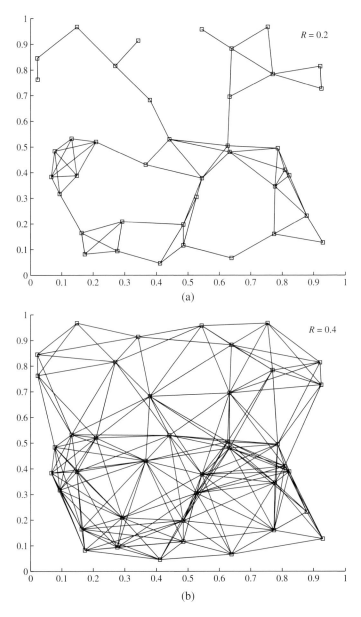

Figure 2.2 Illustration of $G(n, R)$ geometric random graphs: (a) sparse (small R) and (b) dense (large R)

range, there is some number of nodes beyond which there is a high probability that the network obtained is connected. This kind of analysis is relevant for random network deployment, as it provides insights into the minimum density that may be needed to ensure that the network is connected.

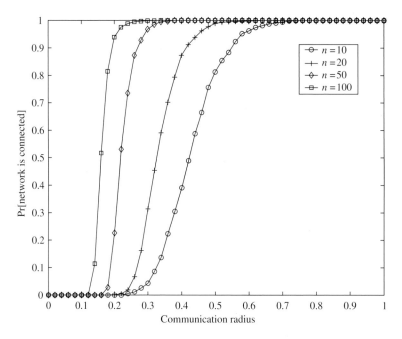

Figure 2.3 Probability of connectivity for a geometric random graph with respect to transmission radius

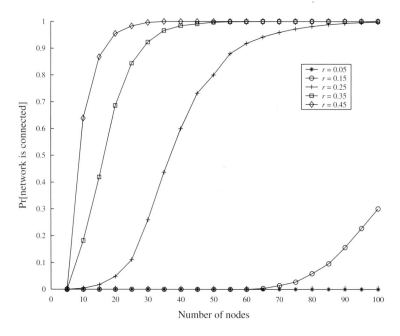

Figure 2.4 Probability of connectivity for a geometric random graph with respect to number of nodes in a unit area

Gupta and Kumar [71] have shown the following result:

Theorem 1
If $\pi R^2 = \frac{\log n + c(n)}{n}$, the network is asymptotically connected almost surely if $\lim\limits_{n \to \infty} c(n) = \infty$ and is disconnected asymptotically almost surely if $\lim\limits_{n \to \infty} c(n) = -\infty$.

In other words, the critical transmission range for connectivity is $O\left(\sqrt{(\frac{\log n}{n})}\right)$. This result is also implied by the work of Penrose [156] on the longest edge of the minimal spanning tree of a random graph. Another surprising result is that the critical radius at which a geometric random graph $G(n, R)$ attains the property that all nodes have at least K neighbors is asymptotically equal to the critical radius at which the graph attains the property of K-connectivity[1] [157].

2.4.2 Monotone properties in G(n, R)

A monotonically increasing property is any graph property that continues to hold if additional edges are added to a graph that already has the property. A graph property is called *monotone* if the property or its inverse are monotonically increasing. Nearly all graph properties of interest from a networking perspective, such as K-connectivity, Hamiltonicity, K-colorability, etc., are monotone. A key theoretical result pertaining to $G(n, R)$ geometric random graphs is that *all monotone properties show critical phase transitions* [65]. Further, all monotone properties are satisfied with high probability within a critical transmission range that is $O\left(\sqrt{(\frac{\log n}{n})} \cdot \log^{1/4} n\right)$.

2.4.3 Connectivity in G(n, K)

Another geometric random graph model is $G(n, K)$, where n nodes are placed at random in a unit area, and each node connects to its K nearest neighbors. This model potentially allows different nodes in the network to use different powers. In this graph, it is known that K must be higher than $0.074 \log n$ and lower than $2.72 \log n$, in order to ensure asymptotically almost sure connectivity [232, 217].

2.4.4 Connectivity and coverage in G$_{grid}$(n, p, R)

Yet another geometric random graph model is the unreliable sensor grid model [191]. In this model n nodes are placed on a square grid within a unit

---

[1] K-connectivity is the property that no $k - 1$ vertices can be removed to disconnect the graph, which, as per Menger's theorem [201], is equivalent to the property that there exist at least K vertex-disjoint paths between all pairs of nodes.

area, p is the probability that a node is active (not failed), and R is the transmission range of each node. For this unreliable sensor grid model, the following properties have been determined:

- For the active nodes to form a connected topology, as well as to cover the unit square region, $p \cdot R^2$ must be $O(\frac{\log n}{n})$.
- The maximum number of hops required to travel from any active node to another is $O\left(\sqrt{(\frac{n}{\log n})}\right)$
- There exists a range of p values sufficiently small such that the active nodes form a connected topology but do not cover the unit square.

2.5 Connectivity using power control

Regardless of whether randomized or structured deployment is performed, once the nodes are in place there is an additional tunable parameter that can be used to adjust the connectivity properties of the deployed network. This parameter is the radio transmission power setting for all nodes in the network.

Power control is quite a complex and challenging cross-layer issue [104]. Increasing radio transmission power has a number of interrelated consequences – some of these are positive, others negative:

- It can extend the communication range, increasing the number of communicating neighboring nodes and improving connectivity in the form of availability of end-to-end paths.
- For existing neighbors, it can improve link quality (in the absence of other interfering traffic).
- It can induce additional interference that reduces capacity and introduces congestion.
- It can cause an increase in the energy expended.

Most of the literature on power-based topology control has been developed for general ad hoc wireless networks, but these results are very much central to the configuration of WSN. We shall discuss some key results and proposed techniques here. Some of these distributed algorithms aim to develop topologies that minimize total power consumption over routing paths, while others aim to minimize transmission power settings of each node (or to minimize the maximum transmission power setting) while ensuring connectivity. These goals are not necessarily complementary; for instance, providing minimum energy paths may require some nodes in the network to have high transmission powers, potentially limiting network lifetime due to partitions caused by rapid battery depletion of these nodes. However, under more dynamic conditions this may not be an issue, as load balancing may be provided through activation of different nodes at different times.

2.5.1 Minimum energy connected network construction (MECN)

Consider the problem of deriving a minimum power network topology for a given deployment of wireless nodes that ensures that the total energy usage for each possible communication path is minimized. A graph topology is defined to be a *minimum power topology*, if for any pair of nodes there exists a path in the graph that consumes the least energy compared with any other possible path. The construction of such a topology is the goal of the the MECN (minimum energy communication network) algorithm [174].

Each node's *enclosure* is defined as the region around it, such that it is always energy-efficient to transmit directly without relaying only for the neighboring nodes within that region. Then the *enclosure graph* is defined as the graph that contains all links between each node and its neighboring nodes in the corresponding enclosure region. The MECN topology control algorithm first constructs the enclosure graph in a distributed manner, then prunes it using a link energy cost-based Bellman–Ford algorithm to determine the minimum power topology.

However, it turns out that the MECN algorithm does not necessarily yield a connected topology with the smallest number of edges. Let $C(u, v)$ be the energy cost for a direct transmission between nodes u and v in the MECN-generated topology. It is possible that there exists another route r between these very nodes, such that the total cost of routing on that path $C(r) < C(u, v)$; in this case the edge (u, v) is redundant.

It has been shown that a topology where no such redundant edges exist is the smallest graph having the *minimum power topology* property [116]. The small minimum energy communication network (SMECN) distributed protocol, while still suboptimal, provides a provably smaller topology with the minimum power property compared to MECN. The advantage of such a topology with a smaller number of edges is primarily a reduced cost for link maintenance.

2.5.2 Minimum common power setting (COMPOW)

The COMPOW protocol [142] ensures that the lowest common power level that ensures maximum network connectivity is selected by all nodes. A number of arguments can be made in favor of using a common power level that is as low as possible (while still providing maximum connectivity) at all nodes: (i) it makes the received signal power on all links symmetric in either direction (although SINR may vary in each direction); (ii) it can provide for an asymptotic network capacity which is quite close to the best capacity achievable without common power levels; (iii) a low common power level provides low-power routes; and (iv) a low power level minimizes contention.

The COMPOW protocol works as follows: first multiple shortest path algorithms (e.g. the distributed Bellman–Ford algorithm) are performed, one at each possible power level. Each node then examines the routing tables generated by the algorithm and picks the lowest power level such that the number of reachable nodes is the same as the number of nodes reachable with the maximum power level.

The COMPOW algorithm can be shown to provide the lowest functional common power level for all nodes in the network while ensuring maximum connectivity, but does suffer from some possible drawbacks. First, it is not very scalable, as each node must maintain a state that is of the order of the number of nodes in the entire network. Further, by strictly enforcing common powers, it is possible that a single relatively isolated node can cause all nodes in the network to have unnecessarily large power levels. Most of the other proposals for topology control with variable power levels do not require common powers on all nodes.

2.5.3 Minimizing maximum power

A work by Ramanathan and Rosales-Hain [168] presents exact (centralized) as well as heuristic (distributed) algorithms that seek to generate a connected topology with non-uniform power levels, such that the maximum power level among all nodes in the network is minimized. They also present algorithms to ensure a biconnected topology, while minimizing the maximum power level. This approach is best suited for the situation where all nodes have the same initial energy level, as it tries to minimize the energy burden on the most loaded device.

2.5.4 Cone-based topology control (CBTC)

The cone-based topology control (CBTC) technique [222, 117] provides a minimal direction-based distributed rule to ensure that the whole network topology is connected, while keeping the power usage of each node as small as possible. The cone-based topology construction is very simple in essence, and involves only a single parameter α, the cone angle. In CBTC each node keeps increasing its transmit power until it has at least one neighboring node in every α cone or it reaches its maximum transmission power limit. It is assumed here that the communication range (within which all nodes are reachable) increases monotonically with transmit power.

The CBTC construction is illustrated in Figure 2.5. On the left we see an intermediate power level for a node at which there exists an α cone in which the node does not have a neighbor. Therefore, as seen on the right, the node must increase its power until at least one neighbor is present in every α.

The original work on CBTC [222] showed that $\alpha \leq 2\pi/3$ suffices to ensure that the network is connected. A tighter result has been obtained [117] that can further reduce the power-level settings at each node:

Theorem 2
If $\alpha \leq 5\pi/6$, then the graph topology generated by CBTC is connected, so long as the original graph, where all nodes transmit at maximum power, is also connected. If $\alpha > 5\pi/6$, then disconnected topologies may result with CBTC.

If the maximum power constraint is ignored so that any node can potentially reach any other node in the network directly with a sufficiently high power setting, then D'Souza *et al.* [41] show that $\alpha = \pi$ is a necessary and sufficient condition for guaranteed network connectivity.

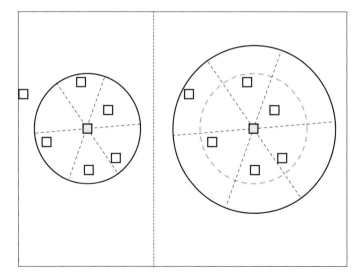

Figure 2.5 Illustration of the cone-based topology control (CBTC) construction

2.5.5 Local minimum spanning tree construction (LMST)

Another approach is to construct a consistent global spanning tree topology in a completely distributed manner [118]. This scheme first runs a local minimum spanning tree (LMST) construction for the portion of the graph that is within visible (max power) range. The local graph is modified with suitable weights to ensure uniqueness, so that all nodes in the network effectively construct consistent LMSTs such that the resultant network topology is connected. The technique ensures that the resulting degree of any node is bounded by 6, and has the property that the topology generated can be pruned to contain only bidirectional links. Simulations have suggested that the technique can outperform both CBTC and MECN in terms of average node degree [118].

2.6 Coverage metrics

Connectivity metrics are generally application independent. In most networks the objective is simply to ensure that there exists a path between every pair of nodes. At most, if robustness is a concern, the K-connectivity (whether there exist K disjoint paths between any pair of nodes) metric may be used. However, the choice of coverage metric is much more diverse and depends highly upon the application.

We shall examine in some detail two qualitatively different sets of coverage metrics that have been considered in several studies: one is the set of *K-coverage metrics* that measure the degree of sensor coverage overlap; the other is the set of *path-observability metrics* that are suitable for applications involving tracking of moving objects.

2.6.1 *K-coverage*

This metric is applicable in contexts where there is some notion of a region being covered by each individual sensor. A field is said to be K-covered if every point in the field is within the overlapping coverage region of at least K sensors. We will limit our discussion here to two dimensions.

Definition 1
Consider an operating region A with n sensor nodes, with each node i providing coverage to a node region $A_i \in A$ (the node regions can overlap). The region A is said to be K-covered if every point $p \in A$ is also in at least K node regions.

At first glance, based on this definition, it may appear that the way to determine that an area is K-covered is to divide the area into a grid of very fine granularity and examine all grid points through exhaustive search to see if they are all K-covered. In an $s \times s$ unit area, with a grid of resolution ϵ unit distance, there will be $(\frac{s}{\epsilon})^2$ such points to examine, which can be computationally intensive. A slightly more sophisticated approach would attempt to enumerate all subregions resulting from the intersection of different sensor node-regions and verify if each of these is K-covered. In the worst case there can be $O(n^2)$ such regions and they are not straightforward to compute. Huang and Tseng [92] prove the interesting result below, which is used to derive an $O(nd \log d)$ distributed algorithm for determining K-coverage.

Definition 2
A sensor is said to be K-perimeter-covered if all points on the perimeter circle of its region are within the perimeters of at least K other sensors.

Theorem 3
The entire region is K-covered if and only if all n sensors are k-perimeter-covered.

These results are shown to hold for the general case when different sensors have different coverage radii. A further improvement on this result is obtained by Wang *et al.* [220]. They prove the following stronger theorem (illustrated in Figure 2.6 for $k = 2$):

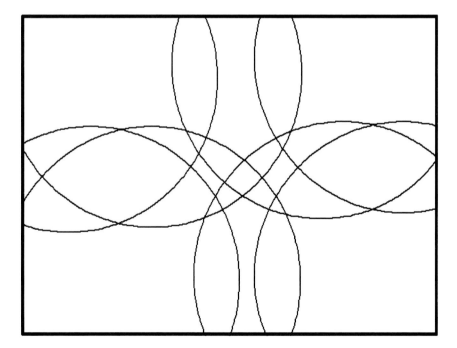

Figure 2.6 An area with 2-coverage (note that all intersection points are 2-covered)

Theorem 4

The entire region is K-covered if and only if all intersection points between the perimeters of the n sensors (and between the perimeter of sensors and the region boundary) are covered by at least K sensors.

Recall that the two main considerations for evaluating a given deployment are coverage and connectivity. Wang *et al.* [220] also provide the following fundamental result pertaining to the relationship between *K*-coverage and *K*-connectivity:

Theorem 5

If a convex region A is K-covered by n sensors with sensing range R_s and communication range R_c, their communication graph is a K-connected network graph so long as $R_c \geq 2R_s$.

2.6.2 Path observation

One class of coverage metrics that has been developed is suitable primarily for tracking targets or other moving objects in the sensor field. A good example of such a metric is the *maximal breach distance* metric [136]. Consider for instance a WSN deployed in a rectangular operational field that a target can traverse from left to right. The maximal breach path is the path that maximizes the distance between the moving target and the nearest sensor during the target's point of nearest approach to any sensor. Intuitively, this metric aims to capture a worst-case notion of coverage, "Given a deployment, how well can an adversary with full knowledge of the deployment avoid observation?"

Given a sensor field, and a set of nodes on it, the maximal breach path is calculated in the following manner:

1. Calculate the Voronoi tessellation of the field with respect to the deployed nodes, and treat it as a graph. A Voronoi tessellation separates the field into separate cells, one for each node, such that all points within each cell are closer to that node than to any other. While the maximal breach path is not unique, it can be shown that at least one maximal breach path must follow Voronoi edges, because they provide points of maximal distance from a set of nodes.
2. Label each Voronoi edge with a cost that represents the minimum distance from any node in the field to that edge.
3. Add a starting (ending) node to the graph to represent the left (right) side of the field, and connect it to all vertices corresponding to intersections between Voronoi edges and the left (right) edge of the field. Label these edges with zero cost.

4. Using a dynamic programming algorithm, determine the path between the starting and ending nodes of the graph that maximizes the lowest-cost edge traversed. This is the maximal breach path. The label of the lowest-cost edge is the maximal breach distance.

An illustration of the maximal breach path can be seen in Figure 2.7(a). Note that there can be several maximal breach paths with the same distance. Given a deployment, the above algorithm can be used to determine the maximal breach distance, which is a worst-case coverage metric. Such an algorithm can be used to evaluate different possible deployments to determine which one provides the best coverage. Note that it is desirable to keep the maximal breach distance as small as possible. Similar to the maximal breach path, there are other possible coverage metrics that try to capture the notion of target observability over a traversal of the field, such as the exposure metric [137] and the lowest probability of detection metric [30].

Unlike the maximal breach distance, which tries to determine the worst-case observability of a traversal by a moving object, the *maximal support distance* [136] aims to provide a best-case coverage metric for moving objects. The maximal support path is the one where the moving node can stay as close as possible to sensor nodes during its traversal of the covered area. Formally, it is the path which tries to minimize the maximum distance between every point on the path and the nearest sensor node. It turns out that the maximal support path can be calculated in a manner similar to the breach path, but this time using the Delaunay triangulation, which connects all nodes in the planar field through line segments that tessellate the field into a set of triangles. Since Delaunay edges

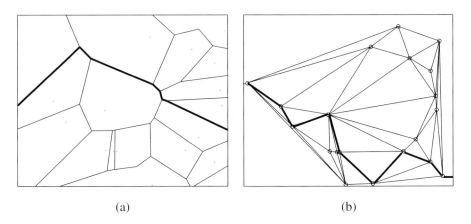

(a)   (b)

Figure 2.7 Illustration of (a) maximal breach path through Voronoi cell edges and (b) minimal support path through Delaunay triangulation edges

represent the shortest way to traverse between any pair of nodes, it can be shown that at least one maximal support path traverses only through Delaunay edges. The edges are labelled with the maximum distance from any point on the edge to the nearest source (i.e. with half the length of the edge). A graph search or dynamic programming algorithm can then be used to find the path through the Delaunay graph (extended to include a start and end node as before) on which the maximum edge cost is minimized. This is illustrated in Figure 2.7(b).

2.6.3 Other metrics

We have focused on two particular kinds of coverage metrics: path-observability metrics and the K-coverage metric. These are by no means representative of all possible coverage metrics. Coverage requirements and metrics can vary a lot from application to application. Some other metrics of interest may be the following:

- Percentage of desired points covered: given a set of desired points in the region where sensor measurements need to be taken, determine the fraction of these within range of the sensors.
- Area of K-coverage: the total area covered by at least K sensors.
- Average coverage overlap: the average number of sensors covering each point in a given region.
- Maximum/average inter-node distance: coverage can also be measured in terms of the maximum or average distance between any pair of nodes.
- Minimum/average probability of detection: given a model of how placement of nodes affects the chances of detecting a target at different locations, the minimum or average of this probability in the area.

2.7 Mobile deployment

We now very briefly touch upon several research efforts that have examined the problems of deployment with mobile nodes. One approach to ensuring non-overlapping coverage with mobile nodes is the use of potential field techniques, whereby the nodes spread out in an area by using virtual repulsive forces to push away from each other [87]. This technique has the great advantage of being completely distributed and localized, and hence easily scales to very large numbers. A similar technique is the distributed self-spreading algorithm (DSSA) [79]. To obtain desirable connectivity guarantees, additional constraints can be incorporated, such as ensuring that each node remains within range of k neighbors [160].

An incremental self-deployment algorithm is described in [88], whereby a new location for placement is calculated at each step based on the current deployment, and the nodes are sequentially shifted so that a new deployment is created with a node moving into that new location and other nodes moving one by one accordingly to fill any gaps. A bidding protocol for deployment of a mixture of mobile and static nodes is described in [219], whereby, after an initial deployment of the static nodes, coverage holes are determined and the mobile nodes move to fill these holes based on bids placed by static nodes. A mutually helpful combination of static sensor nodes and mobile nodes is described in [8], where a robotic node's mobile explorations help determine where static nodes are to be deployed, and the deployed static node then provides guidance to the robot's exploration. The deployment of static sensor nodes from an autonomous helicopter is described in [33], where the sensor nodes are first dropped from the air and self-configure to determine their connectivity. If the network is found to be disconnected, the helicopter is informed about where to deploy additional nodes.

2.8 Summary

We observe that the deployment of a sensor network can have a significant impact on its operational performance and therefore requires careful planning and design. The fundamental objective is to ensure that the network will have the desired connectivity and application-specific coverage properties during its operational lifetime. The two major methodologies for deployment are: (a) structured placement and (b) random scattering of nodes. Particularly for small–medium-scale deployments, where there are equipment cost constraints and a well-specified set of desired sensor locations, structured placements are desirable. In other applications involving large-scale deployments of thousands of inexpensive nodes, such as surveillance of remote environments, a random scattering of nodes may be the most flexible and convenient option. Nodes may be over deployed, with redundancy for reasons of robustness, or else deployed/replaced incrementally as nodes fail.

Geometric random graphs offer a useful methodology for analyzing and determining density and parameter settings for random deployments of WSN. There exist several geometric random graph models including $G(n, R)$, $G(n, K)$, $G_{grid}(n, p, R)$. One common feature of all these models is that asymptotically the condition to ensure connectivity is that each node have $O(\log n)$ neighbors on average. All monotone properties (including most coverage and coverage properties of interest) in $G(n, R)$ are known to undergo sharp phase transitions at critical thresholds that pertain to points of resource-efficient system operation.

Once the nodes have been placed, the connectivity properties of the network can be adjusted by modification of the transmit powers of nodes. Many distributed algorithms have been developed for variable power-based topology control. Power control techniques must provide connectivity, while taking into account diverse factors, including interference minimization and energy reduction.

Coverage metrics are particularly application dependent. Two classes of metrics that have been studied by several researchers are the K-coverage and path-observability metrics. A fundamental theoretical result tying coverage to connectivity is that, so long as the connectivity and sensing ranges satisfy the condition $R_c \geq 2R_s$, K-coverage implies K-connectivity.

The deployment of mobile robotic sensor nodes is important for some applications, and raises related challenges. Distributed potential-based approaches appear particularly promising for autonomous mobile deployment.

Exercises

2.1 *Topology selection:* Consider a remote deployment consisting of three sensor nodes A, B, C, and a gateway node D. The following set of stationary packet reception probabilities (i.e. the probability that a packet is received successfully) has been determined for each link from experimental measurements: [A–B: 0.65, A–C: 0.95, A–D: 0.95, B–A: 0.90, B–C: 0.3, B–D: 0.99, C–A: 0.95, C–B: 0.6, C–D: 0.3]. Assuming all traffic must originate at the sources (A, B, C) and end at the gateway (D), explain why a single-hop star topology is unsuitable for this deployment, and suggest a topology that would be more suitable.

2.2 *The $G(n, R)$ geometric random graph:* In this question assume all nodes are deployed randomly with a uniform distribution in a unit square area. Determine the following through simulations:

   (a) Estimate the probability of connectivity when $n = 40$, $R = 0.20$.
   (b) Estimate the minimum number of nodes n_{min} that need to be deployed to guarantee network connectivity with greater than 80% probability if $R = 0.2$.
   (c) Plot, with respect to R, the probability that each node has at least K neighbors for $k = 1$, 2, and 3, assuming $n = 100$.

(d) Plot the probability of network connectivity for different values of n and R, as in Figure 2.3, but with respect to a different normalized x-axis $R \cdot \sqrt{(\frac{n}{\log n})}$. What do you observe?

2.3 *The $G(n, K)$ geometric random graph:* Through simulations of $G(n, K)$ try to estimate the numerical value of the ratio $\frac{K_{crit}}{\log n}$ for large n, where K_{crit} is the minimum number of neighbors each node must have to ensure connectivity. How does this relate to the bounds described in Section 2.4.3?

2.4 *Enclosure region:* Assume a 1-D deployment along the x-axis in which a node A is located on the far left corner at $x=0$, and there is a neighboring node B a unit distance to its right located at $x=1$. Let the transmission energy cost of a single packet sent directly from one node to another node at distance d be given as $a+b \cdot d^2$. Derive an expression for the coordinate point x_{min} such that only for nodes located with coordinates greater than x_{min} is it energy-efficient for a packet from A to be routed through B, rather than through a direct transmission.

2.5 *Isolated nodes with common power:* Consider a network of five nodes located at the following coordinate points A at $(0,0)$, B at $(0,1)$, C at $(1,0)$, D at $(1,1)$, E at $(5,0)$. Assume the transmission power necessary to reach a node at distance d is again $a+b \cdot d^2$. What is the common transmission power necessary to form a connected network topology consisting of all nodes? What is the common transmission power necessary to form a connected topology if node E can be left out of the network? Does this suggest a possible drawback of the basic COMPOW protocol? How could it be fixed?

2.6 *Cone-based topology construction:* Consider a network of 100 nodes laid out on a square grid at points $(m/10, n/10)$, where m and n are each any integer between 0 and 10. How many neighbors does each node have with the CBTC construction if $\alpha = \pi/3$? What is the final transmission range setting of each node in this case?

2.7 *K-coverage:* Consider the square region from $(0,0)$ to $(1,1)$ with 100 sensor nodes again located on the grid at coordinate points $(m/10, n/10)$. Assume all nodes have the same sensing range R_s. What should R_s be in order to ensure that the area is K-covered, for $k=1, 2, 4$? For these cases, give a setting of the communication range R_c that will also ensure K-connectivity.

2.8   *Path observation metrics:* Consider a square area and a target/mobile node
that is always known to enter from the left side and leave through the right
side after traversing a linear trajectory.

(a) Assume the target can enter from any point on the left side and leave
from any point on the right side. How should a given number of nodes
be deployed to ensure that the worst-case breach distance (the point
of nearest approach to any sensor node) is minimized?

(b) Assume the mobile node always enters from the exact middle of the
left side and leaves from the exact middle of the right side. How should
a given number of nodes be deployed to give the best support path
(such that the maximum distance between any point on the path and
the nearest sensor node is minimized)?

Localization

3.1 Overview

Wireless sensor networks are fundamentally intended to provide information about the spatio-temporal characteristics of the observed physical world. Each individual sensor observation can be characterized essentially as a tuple of the form $< S, T, M >$, where S is the spatial location of the measurement, T the time of the measurement, and M the measurement itself. We shall address the following fundamental question in this chapter: How can the spatial location of nodes be determined?

The location information of nodes in the network is fundamental for a number of reasons:

1. **To provide location stamps** for individual sensor measurements that are being gathered.
2. **To locate and track point objects** in the environment.
3. **To monitor the spatial evolution of a diffuse phenomenon** over time, such as an expanding chemical plume. For instance, this information is necessary for in-network processing algorithms that determine and track the changing boundaries of such a phenomenon.
4. **To determine the quality of coverage**. If node locations are known, the network can keep track of the extent of spatial coverage provided by active sensors at any time.
5. **To achieve load balancing** in topology control mechanisms. If nodes are densely deployed, geographic information of nodes can be used to selectively shut down some percentage of nodes in each geographic area to conserve energy, and rotate these over time to achieve load balancing.

6. **To form clusters**. Location information can be used to define a partition of
   the network into separate clusters for hierarchical routing and collaborative
   processing.
7. **To facilitate routing** of information through the network. There are a number
   of geographic routing algorithms that utilize location information instead of
   node addresses to provide efficient routing.
8. **To perform efficient spatial querying**. A sink or gateway node can issue
   queries for information about specific locations or geographic regions. Location information can be used to scope the query propagation instead of
   flooding the whole network, which would be wasteful of energy.

We should, at the outset, make it clear that localization may not be a significant
challenge in all WSN. In structured, carefully deployed WSN (for instance in
industrial settings, or scientific experiments), the location of each sensor may be
recorded and mapped to a node ID at deployment time. In other contexts, it may
be possible to obtain location information using existing infrastructure, such as
the satellite-based GPS [141] or cellular phone positioning techniques [218].

However, these are not satisfactory solutions to all contexts. A-priori knowledge of sensor locations will not be available in large-scale and ad hoc deployments. A pure-GPS solution is viable only if all nodes in the network can be
provided with a potentially expensive GPS receiver and if the deployed area
provides good satellite coverage. Positioning using signals directly from cellular
systems will not be applicable for densely deployed WSN, because they generally
offer poor location accuracy (on the order of tens of meters). If only a subset of
the nodes have known location a priori, the position of other nodes must still be
determined through some localization technique.

3.2 Key issues

Localization is quite a broad problem domain [80, 185], and the component
issues and techniques can be classified on the basis of a number of key questions.

1. **What to localize?** This refers to identifying which nodes have a priori
   known locations (called reference nodes) and which nodes do not (called
   unknown nodes). There are a number of possibilities. The number and fraction of reference nodes in a network of n nodes may vary all the way
   from 0 to $n - 1$. The reference nodes could be static or mobile; as could

the unknown nodes. The unknown nodes may be cooperative (e.g. partic-
ipants in the network, or robots traversing the networked area) or non-
cooperative (e.g. targets being surveilled). The last distinction is important
because non-cooperative nodes cannot participate actively in the localization
algorithm.

2. **When to localize?** In most cases, the location information is needed for
   all unknown nodes at the very beginning of network operation. In static
   environments, network localization may thus be a one-shot process. In other
   cases, it may be necessary to provide localization on-the-fly, or refresh the
   localization process as objects and network nodes move around, or improve
   the localization by incorporating additional information over time. The time
   scales involved may vary considerably from being of the order of minutes to
   days, even months.

3. **How well to localize?** This pertains to the resolution of location information
   desired. Depending on the application, it may be required for the localization
   technique to provide absolute (x, y, z) coordinates, or perhaps it will suffice
   to provide relative coordinates (e.g. "south of node 24 and east of node 22");
   or symbolic locations (e.g. "in room A", "in sector 23", "near node 21").
   Even in case of absolute locations, the required accuracy may be quite different (e.g. as good as ± 20 cm or as rough as ± 10 m). The technique must
   provide the desired type and accuracy of localization, taking into account the
   available resources (such as computational resources, time-synchronization
   capability, etc.).

4. **Where to localize?** The actual location computation can be performed at
   several different points in the network: at a central location once all component
   information such as inter-node range estimates is collected; in a distributed
   iterative manner within reference nodes in the network; or in a distributed
   manner within unknown nodes. The choice may be determined by several
   factors: the resource constraints on various nodes, whether the node being
   localized is cooperative, the localization technique employed, and, finally,
   security considerations.

5. **How to localize?** Finally, different signal measurements can be used as
   inputs to different localization techniques. The signals used can vary from
   narrowband radio signal strength readings or packet-loss statistics, UWB RF
   signals, acoustic/ultrasound signals, infrared. The signals may be emitted and
   measured by the reference nodes, by the unknown nodes, or both. The basic
   localization algorithm may be based on a number of techniques, such as
   proximity, calculation of centroids, constraints, ranging, angulation, pattern
   recognition, multi-dimensional scaling, and potential methods.

3.3 Localization approaches

Generally speaking, there are two approaches to localization:

1. **Coarse-grained localization using minimal information:** These typically use a small set of discrete measurements, such as the information used to compute location. Minimal information could include binary proximity (can two nodes hear each other or not?), near–far information (which of two nodes is closer to a given third node?), or cardinal direction information (is one node in the north, east, west, or south sector of the other given node?).
2. **Fine-grained localization using detailed information:** These are typically based on measurements, such as RF power, signal waveform, time stamps, etc., that are either real-valued or discrete with a large number of quantization levels. These include techniques based on radio signal strengths, timing information, and angulation.

The tradeoff that emerges between the two approaches is easy to see: while minimal information techniques are simpler to implement, and likely involve lower resource consumption and equipment costs, they provide lower accuracy than the detailed information techniques. We shall now describe specific techniques in detail.

We shall start first with the *node localization* problem involving a single unknown node and several reference nodes, and then discuss the problem of *network localization* where there are several unknown nodes in a multi-hop network.

3.4 Coarse-grained node localization using minimal information

3.4.1 Binary proximity

Perhaps the most basic location technique is that of binary proximity – involving a simple decision of whether two nodes are within reception range of each other. A set of references nodes are placed in the environment in some non-overlapping (or nearly non-overlapping) manner. Either the reference nodes periodically emit beacons, or the unknown node transmits a beacon when it needs to be localized. If reference nodes emit beacons, these include their location IDs. The unknown node must then determine which node it is closest to, and this provides a coarse-grained localization. Alternatively, if the unknown node emits a beacon, the reference node that hears the beacon uses its own location to determine the location of the unknown node.

An excellent example of proximity detection as a means for localization is the Active Badge location system [221] meant for an indoor office environment. This system consists of small badge cards (about 5 square centimeters in size and less than a centimeter thick) sending unique beacon signals once every 15 seconds with a 6 meter range. The active badges, in conjunction with a wired sensor network that provides coverage throughout a building, provide room-level location resolution. A much larger application of localization, using binary proximity detection, is with passive radio frequency identification (RFID) tags, which can be detected by readers within a similar short range [52]. Today there are a large number of inventory-tracking applications envisioned for RFIDs. A key difference in RFID proximity detection compared with active badges is that the unknown nodes are passive tags, being queried by the reference nodes in the sensor network. These examples show that even the simplest localization technique can be of considerable use in practice.

3.4.2 Centroid calculation

The same proximity information can be used to greater advantage when the density of reference nodes is sufficiently high that there are several reference nodes within the range of the unknown node. Consider a two-dimensional scenario. Let there be n reference nodes detected within the proximity of the unknown node, with the location of the ith such reference denoted by (x_i, y_i). Then, in this technique, the location of the unknown node (x_u, y_u) is determined as

$$x_u = \frac{1}{n} \sum_{i=1}^{n} x_i$$

$$y_u = \frac{1}{n} \sum_{i=1}^{n} y_i \qquad (3.1)$$

This simple centroid technique has been investigated using a model with each node having a simple circular range R in an infinite square mesh of reference nodes spaced a distance d apart [16]. It is shown through simulations that, as the overlap ratio R/d is increased from 1 to 4, the average RMS error in localization is reduced from $0.5d$ to $0.25d$.

3.4.3 Geometric constraints

If the bounds on radio or other signal coverage for a given node can be described by a geometric shape, this can be used to provide location estimates by determining which geometric regions that node is constrained to be in, because of intersections between overlapping coverage regions.

For instance, the region of radio coverage may be upper-bounded by a circle of radius R_{max}. In other words, if node B hears node A, it knows that it must be no more than a distance R_{max} from A. Now, if an unknown node hears from several reference nodes, it can determine that it must lie in the geometric region described by the intersection of circles of radius R_{max} centered on these nodes. This can be extended to other scenarios. For instance when both lower R_{min} and upper bounds R_{max} can be determined, based on the received signal strength, the shape for a single node's coverage is an annulus; when an angular sector $(\theta_{min}, \theta_{max})$ and a maximum range R_{max} can be determined, the shape for a single node's coverage would be a cone with given angle and radius.

Although arbitrary shapes can be potentially computed in this manner, a computational simplification that can be used to determine this bounded region is to use rectangular bounding boxes as location estimates. Thus the unknown node determines bounds $x_{min}, y_{min}, x_{max}, y_{max}$ on its position.

Figure 3.1 illustrates the use of intersecting geometric constraints for localization. Localization techniques using such geometric regions were first described by Doherty *et al.* [40]. One of the nice features of these techniques is that not only can the unknown nodes use the centroid of the overlapping region as a specific location estimate if necessary, but they can also determine a bound on the location error using the size of this region.

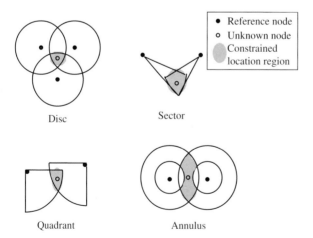

Figure 3.1 Localization using intersection of geometric constraints

When the upper bounds on these regions are tight, the accuracy of this geometric approach can be further enhanced by incorporating "negative information"
about which reference nodes are *not* within range [54].

3.4.4 Approximate point in triangle (APIT)

A related approach to localization using geometric constraints is the approximate
point-in-triangle (APIT) technique [72]. APIT is similar to the above techniques
in that it provides location estimates as the centroid of an intersection of regions.
Its novelty lies in how the regions are defined – as triangles between different
sets of three reference nodes (rather than the coverage of a single node). This
is illustrated in Figure 3.2. It turns out that an exact determination of whether
an unknown node lies within the triangle formed by three reference nodes is
impossible if nodes are static because wireless signal propagation is non-ideal.
An approximate solution can be determined using near–far information [72], i.e.
the ability to determine which of two nodes is nearer a third node based on signal
reception. One caveat for the APIT technique is that it can provide erroneous
results, because the determination of whether a node lies within a particular
triangle requires quite a high density of nodes in order to provide good location
accuracy.

3.4.5 Identifying codes

There is another interesting technique that utilizes overlapping coverage regions
to provide localization. In this technique, referred to as the identifying code

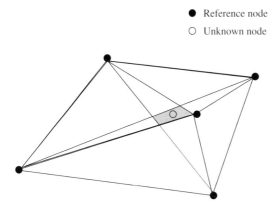

Figure 3.2 The approximate point-in-triange (APIT) technique

construction (ID-CODE) algorithm [173], the sensor deployment is planned in such a way as to ensure that each resolvable location is covered by a unique set of sensors.

The algorithm runs on a deployment region graph $G = (V, E)$ in which vertices V represent the different regions, and the edges E represent radio connectivity between regions. Let $B(v)$ be the set of vertices that are adjacent to v, together with v itself. A set of vertices $C \in V$ is referred to as an *identifying code*, if, for all $u, v \in V$, $B(v) \cap C \neq B(u) \cap C$.

It can be shown that a graph is distinguishable, i.e. there exists an identifying code for it, if and only if there are no two vertices u, v such that $B(u) = B(v)$. The goal of the algorithm is to construct an identifying code for any distinguishable graph, with each vertex in the code corresponding to a region where a reference node must be placed. Once this is done, by the definition of the identifying code, each location region in the graph will be covered by a unique set of reference nodes. This is illustrated in Figure 3.3.

While the entire set of vertices V itself is an identifying code, such a placement of a reference node in each region would clearly be inefficient. On the other hand, obtaining a minimal cardinality identifying code is known to be NP-complete. The algorithm ID-CODE is a polynomial greedy heuristic that provides good solutions in practice. There also exists a robust variant of this algorithm called r-ID-CODE [173] that can provide robust identification, i.e. guaranteeing a unique set of IDs for each location, even if there is addition or deletion of up to r ID values.

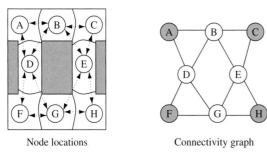

Node locations   Connectivity graph

Transmitters A, F, C, H provide unique IDs for all node locations

V:	A	B	C	D	E	F	G	H
ID:	A	A,C	C	A,F	C,H	F	F,H	H

Figure 3.3 Illustration of the ID-CODE technique showing uniquely identifiable regions

3.5 Fine-grained node localization using detailed information

We now examine techniques based on detailed information. These include tri-angulation using distance estimates, pattern matching, and sequence decoding. Although used in the large-scale GPS, basic time-of-flight techniques using RF signals are not capable of providing precise distance estimates over short ranges typical of WSN because of synchronization limitations. Therefore other tech-niques such as radio signal strength (RSS) measurements and time difference of arrival (TDoA) must be used for distance-estimation.

3.5.1 Radio signal-based distance-estimation (RSS)

To a first-order approximation, mean radio signal strengths diminish with distance according to a power law. One model that is used for wireless radio propagation is the following [171]:

$$P_{r,dB}(d) = P_{r,dB}(d_0) - \eta 10 \log\left(\frac{d}{d_0}\right) + X_{\sigma,dB} \qquad (3.2)$$

where $P_{r,dB}(d)$ is the received power at distance d and $P(d_0)$ is the received power at some reference distance d_0, η the path-loss exponent, and $X_{\sigma,dB}$ a log-normal random variable with variance σ^2 that accounts for fading effects. So, in theory, if the path-loss exponent for a given environment is known the received signal strength can be used to estimate the distance. However, the fading term often has a large variance, which can significantly impact the quality of the range estimates. This is the reason RF-RSS-based ranging techniques may offer location accuracy only on the order of meters or more [154]. RSS-based ranging may perform much better in situations where the fading effects can be combatted by diversity techniques that take advantage of separate spatio-temporally correlated signal samples.

3.5.2 Distance-estimation using time differences (TDoA)

As we have seen, time-of-flight techniques show poor performance due to preci-sion constraints, and RSS techniques, although somewhat better, are still limited by fading effects. A more promising technique is the combined use of ultra-sound/acoustic and radio signals to estimate distances by determining the TDoA of these signals [164, 183, 223]. This technique is conceptually quite simple, and is illustrated in Figure 3.4. The idea is to simultaneously transmit both the radio and acoustic signals (audible or ultrasound) and measure the times T_r and T_s of the

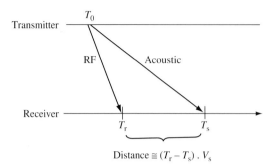

Figure 3.4 Ranging based on time difference of arrival

arrival of these signals respectively at the receiver. Since the speed of the radio signal is much larger than the speed of the acoustic signal, the distance is then simply
estimated as $(T_s - T_r) \cdot V_s$, where V_s is the speed of the acoustic signal.

One minor limitation of acoustic ranging is that it generally requires the nodes
to be in fairly close proximity to each other (within a few meters) and preferably
in line of sight. There is also some uncertainty in the calculation because the
speed of sound varies depending on many factors such as altitude, humididity,
and air temperature. Acoustic signals also show multi-path propagation effects
that may impact the accuracy of signal detection. These can be mitigated to a
large extent using simple spread-spectrum techniques, such as those described in
[61]. The basic idea is to send a pseudo-random noise sequence as the acoustic
signal and use a matched filter for detection, (instead of using a simple chirp and
threshold detection).

On the whole, acoustic TDoA ranging techniques can be very accurate in practical settings. For instance, it is claimed in [183] that distance can be estimated
to within a few centimeters for node separations under 3 meters. Of course,
the tradeoff is that sensor nodes must be equipped with acoustic transceivers in
addition to RF transceivers.

3.5.3 Triangulation using distance estimates

The location of the unknown node (x_0, y_0) can be determined based on measured
distance estimates $\hat{d}_i$ to n reference nodes $\{(x_1, y_1), \ldots, (x_i, y_i), \ldots, (x_n, y_n)\}$.
This can be formulated as a least squares minimization problem.

Let d_i be the correct Euclidean distance to the n reference nodes, i.e.:

$$d_i = \sqrt{(x_i - x_0)^2 + (y_i - y_0)^2} \tag{3.3}$$

Thus the difference between the measured and actual distances can be represented as

$$\rho_i = \hat{d}_i - d_i \tag{3.4}$$

The least squares minimization problem is then to determine the (x_0, y_0) that minimizes $\sum_{i=1}^{n}(\rho_i)^2$. This problem can be solved by the use of gradient descent techniques or by iterative successive approximation techniques such as described in [146]. An alternative is the following approach, which provides a numerical solution to an over-determined $(n \geq 3)$ linear system [183].

The over-determined linear system can be obtained as follows. Rearranging and squaring terms in equation (3.3), we would have n such equations:

$$x_i^2 + y_i^2 - d_i^2 = 2x_0 x_i + 2y_0 y_i - (x_0^2 + y_0^2) \tag{3.5}$$

By subtracting out the nth equation from the rest, we would have $n - 1$ equations of the following form:

$$x_i^2 + y_i^2 - x_n^2 - y_n^2 - d_i^2 + d_n^2 = x_0 2(x_i - x_n) + y_0 2(y_i - y_n) \tag{3.6}$$

which yields the linear relationship

$$\mathbf{A}\overline{x} = \mathbf{B} \tag{3.7}$$

where $\mathbf{A}$ is an $(n - 1) \times 2$ matrix, such that the ith row of $\mathbf{A}$ is $[2(x_i - x_n) \quad 2(y_i - y_n)]$, $\overline{x}$ is the column vector representing the coordinates of the unknown location $[x_0 \quad y_0]^T$, and $\mathbf{B}$ is the $(n - 1)$ element column vector whose ith term is the expression $x_i^2 + y_i^2 - x_n^2 - y_n^2 - d_i^2 + d_n^2$. Now, in practice we cannot determine $\mathbf{B}$, since we have access to only the estimated distances, so we can calculate instead the elements of the related vector $\hat{\mathbf{B}}$, which is the same as $\mathbf{B}$ with $\hat{d}_i$ substituted for d_i. Now the least squares solution to equation (3.7) is to determine an estimate for $\overline{x}$ that minimizes $\|\mathbf{A}\overline{x} - \hat{\mathbf{B}}\|_2$. Such an estimate is provided by

$$\overline{x} = (\mathbf{A}^T \mathbf{A})^{-1} \mathbf{A}^T \hat{\mathbf{B}} \tag{3.8}$$

Solving for the above may not directly yield a numerical solution if the matrix $\mathbf{A}$ is ill-conditioned, so a recommended approach is to instead use the *pseudo-inverse* $\mathbf{A}^+$ of the matrix $\mathbf{A}$:

$$\overline{x} = \mathbf{A}^+ \hat{\mathbf{B}} \tag{3.9}$$

3.5.4 Angle of arrival (AoA)

Another possibility for localization is the use of angular estimates instead of distance estimates. Angles can potentially be estimated by using rotating directional beacons, or by using nodes equipped with a phased array of RF or ultrasonic receivers. A very simple localization technique, involving three rotating reference beacons at the boundary of a sensor network providing localization for all interior nodes, is described in [143]. A more detailed description of AoA-based triangulation techniques is provided in [147].

Angulation with ranging is a particularly powerful combination [27]. In theory, if the angular information provided to a given reference node can be combined with a good distance estimate to that reference node, then localization can be performed with a single reference using polar coordinate transformation. While the accuracy and precision with which angles in real systems can be determined are unclear, significant improvements can be obtained by combining accurate ranging estimates with even coarse-grained angle estimates.

3.5.5 Pattern matching (RADAR)

An alternative to measuring distances or angles that is possible in some contexts is to use a pre-determined "map" of signal coverage in different locations of the environment, and use this map to determine where a particular node is located by performing pattern matching on its measurements. An example of this technique is RADAR [5]. This technique requires the prior collection of empirical measurements (or high-fidelity simulation model) of signal strength statistics (mean, variance, median) from different reference transmitters at various locations. It is also important to take into account the directional orientation of the receiving node, as this can result in significant variations. Once this information is collected, any node in the area is localized by comparing its measurements from these references to determine which location matches the received pattern best. This technique has some advantages, in particular as a pure RF technique it has the potential to perform better than the RSS-based distance-estimation and the triangulation approach we discussed before. However, the key drawback of the technique is that it is very location specific and requires intensive data collection prior to operation; also it may not be useful in settings where the radio characteristics of the environment are highly dynamic.

3.5.6 RF sequence decoding (ecolocation)

The ecolocation technique [238] uses the relative ordering of received radio signal strengths for different references as the basis for localization. It works as follows:

1. The unknown node broadcasts a localization packet.
2. Multiple references record their RSSI reading for this packet and report it to a common calculation node.
3. The multiple RSSI readings are used to determine the ordered sequence of references from highest to lowest RSSI.
4. The region is scanned for the location for which the correct ordering of references (as measured by Euclidean distances) has the "best match" to the measured sequence. This is considered the location of the unknown node.

In an ideal environment, the measured sequence would be error free, and ecolocation would return the correct location region. However, in real environments, because of multi-path fading effects, the measured sequence is likely to be corrupted with errors. Some references, which are closer than others to the true location of the unknown node, may show a lower RSSI, while others, which are farther away, may appear earlier in the sequence. Therefore the sequence must be decoded in the presence of errors. This is why a notion of "best match" is needed.

The best match is quantified by deriving the $n(n-1)/2$ pair-wise ordering constraints (e.g. reference A is closer than reference B, reference B is closer than reference C, etc.) at each location, and determining how many of these constraints are satisfied/violated in the measured sequence. The location which provides the maximum number of satisfied constraints is the best match. Simulations and experiments suggest that ecolocation can provide generally more accurate localizations compared with other RF-only schemes, including triangulation using distance estimates. Intuitively, this is because the ordered relative sequence of RSSI values at the references provides robustness to fluctuations in the absolute RSSI value.

3.6 Network-wide localization

3.6.1 Issues

So far, we have focused on the problem of *node localization*, which is that of determining the location of a single unknown node given a number of nearby references. A broader problem in sensor systems is that of *network localization*, where several unknown nodes have to be localized in a network with a few reference nodes. While the network localization problem can rarely be neatly decomposed into a number of separate node localization problems (since there may be unknown nodes that have no reference nodes within range), the node localization and ranging techniques described above do often form an integral component of solutions to network localization.

The performance of network localization depends very much on the resources and information available within the network. Several scenarios are possible: for instance there may be no reference nodes at all, so that perhaps only relative coordinates can be determined for the unknown nodes; if present, the number/density of reference nodes may vary (generally the more reference nodes there are, the lower the network localization error); there may be just a single mobile reference. Information about which nodes are within range of each other may be available; or inter-node distance estimates may be available; inter-node angle information may be available.

Some network localization approaches are centralized, in which all the available information about known nodes, and the inter-node distances or other inter-node relationships are provided to a central node, where the solution is computed. Such a centralized approach may be sufficient in moderate-sized networks, where the nodes in the network need to be localized only once, post-deployment. Other network localization approaches are distributed, often involving the iterative communication of updated location information.

There may be several ways to measure the performance of network localization. If the ground truth is available, these can range from the full distribution/histogram of location errors, to the mean location error, to the percentage of unknown nodes that can be located within a desired accuracy. Alternatively some localization approaches provide an inherent way to estimate the uncertainty associated with each node's calculated location.

3.6.2 Constraint-based approaches

Geometric constraints can often be expressed in the form of linear matrix inequalities and linear constraints [40]. This applies radial constraints (two nodes are determined to be within range R of each other), annular constraints (a node is determined to be within ranges $[R_{min}, R_{max}]$ of another), angular constraints (a node is determined to be within a particular angular sector of another), as well as other convex constraints. Information about a set of reference nodes together with these constraints (which provide the inter-node relationships amongst reference as well as unknown nodes) describes a feasible set of constraints for a semidefinite program. By selecting an appropriate objective function for the program, the constraining rectangle, which bounds the location for each unknown node, can be determined.

When using bounding rectangles, a distributed iterative solution can be used [54]. In this solution, at each step nodes broadcast to their neighbors their current constrained region, which is calculated based on the overheard information about their neighbors' constrained regions at the previous step. If continued for a sufficient number of iterations, or until there is no longer a significant improvement in the bounds, this can provide a solution that is near or at optimal.

Network localization can also be performed in the presence of mobile reference/target nodes [54]. If the mobile node is a reference and able to provide an accurate location beacon, then it can substantially improve localization over time, because each new observation of the moving beacon introduces additional constraints. In theory the location error can be reduced to an arbitrarily small quantity if the moving beacon is equally likely to move to any point in the network. If the mobile node is a non-cooperative target, then the distributed iterative algorithm can be extended to provide simultaneous network localization and tracking with performance that improves over time.

3.6.3 RSS-based joint estimation

If radio signal strengths can be measured between all pairs of nodes in the network that are within detection range, then a joint maximum likelihood estimation (MLE) technique can be used to determine the location of unknown nodes in a network [154]. In the joint MLE technique, first an expression is derived for the likelihood that the obtained matrix of power measurements would be received given a particular location set for all nodes; the objective is then to find the location set that maximizes this likelihood. The performance of this joint MLE technique has been verified through simulations and experiments to show that localization of the order of 2 meters is possible when there is a high density of unknown nodes, even if there are only a few reference nodes sparsely placed.

3.6.4 Iterative multilateration

The iterative multilateration technique [183] is applicable whenever inter-node distance information is available between all neighboring nodes (regardless of whether it is obtained through RSS measurements or TDoA or any other approach). The algorithm is quite simple. It applies the basic triangulation technique for node localization (see Section 3.5.3 above) in an iterative manner to determine the locations of all nodes. One begins by determining the location of an unknown node that has the most reference nodes in its neighborhood. In a distributed version, the location of any node with sufficient references in its neighborhood may be calculated as the initial step. This node is then added to the set of reference nodes and the process is repeated. Figure 3.5 shows an example of a network with one possible sequence in which unknown nodes can each compute their location so long as at least three of their neighbors have known or already computed locations.

Note that a version of iterative multilateration can also be utilized if only connectivity information is available. In such a case, a centroid calculation could be used at each iterative step by the unknown nodes, instead of using distance-based triangulation.

The iterative multilateration technique suffers from two shortcomings: first, it may not be applicable if there is no node that has sufficient (≥ 3 for the 2D plane) reference nodes in its neighborhood; second, the use of localized unknown nodes as reference nodes can introduce substantial cumulative error in the network localization (even if the more certain high-reference neighborhood nodes are used earlier in the iterative process).

3.6.5 Collaborative multilateration

One approach to tackle the deficiency of iterative multilateration with respect to nodes with insufficient reference neighbors is the collaborative multilateration

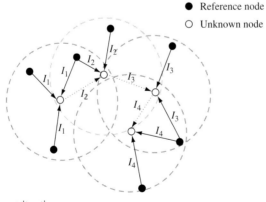

● Reference node

○ Unknown node

$I_n = n$th iteration.

Figure 3.5 Illustration of sequence of iterative multilateration steps

approach described by Savvides *et al.* [183]. The key insight is to determine collaborative subgraphs within the network that contain reference and unknown nodes in a topology such that their positions and inter-node distances can be written as an over-constrained set of quadratic equations with a unique solution for the location of unknown nodes (which can be obtained through gradient descent or local search algorithms). Used in conjunction with iterative multilateration, this technique is generally useful in portions of the network where the reference node density is low.

3.6.6 Multi-hop distance-estimation approaches

An alternative approach to network localization utilizes estimates of distances to reference nodes that may be several hops away [146]. These distances are propagated from reference nodes to unknown nodes using a basic distance-vector technique. There are three variants of this approach:

1. DV-hop: In this approach, each unknown node determines its distance from various reference nodes by multiplying the least number of hops to the reference nodes with an estimated average distance per hop. The average distance per hop depends upon the network density, and is assumed to be known.
2. DV distance: If inter-node distance estimates are directly available for each link in the graph, then the distance-vector algorithm is used to determine the distance corresponding to the shortest distance path between the unknown nodes and reference nodes.
3. Euclidean propagation: Geometric relations can be used in addition to distance estimates to determine more accurate estimates to reference nodes. For instance consider a quadrilateral ABCR, where A and R are at opposite ends; if node A knows the distances AB, AC, BC and nodes B and C have estimates of their distance to the reference R, then A can use the geometric relations inherent in this quadrilateral to calculate an estimated distance to R.

Once distance estimates are available from each unknown node to different reference nodes throughout the network, a triangulation technique (such as described in Section 3.5.3) can be employed to determine their locations. Through simulations, it has been seen that location errors for most nodes can be kept within a typical single-hop distance [146]. In a comparative simulation study of these approaches, it has been shown that the relative performance of these three schemes depends on factors such as the radio range and accuracy of available distance estimates [113].

3.6.7 Refinement

Once a possible initial estimate for the location of unknown nodes has been determined through iterative multilateration/collaborative multilateration or the distance-vector estimation approaches, additional refinement steps can be applied [182]. Each node continues to iterate, obtaining its neighbors location estimates and using them to calculate an updated location using triangulation. After some iterations, the position updates become small and this refinement process can be stopped.

3.6.8 Force-calculation approach

An inherently distributed iterative approach to network localization is to use a physics-based analogy [86, 214]. Each node first picks a reasonable initial guess as to its location, which need not be very accurate. If $d_{i,j}$ is the calculated distance between the two nodes as per their current positions, $\hat{d}_{i,j}$ the estimated distance and $\overrightarrow{u_{i,j}}$ the unit vector between them, and H_i is the set of all neighboring nodes of i, then a vector force on a link and the resultant force on a node can be respectively defined as

$$\overrightarrow{F_{i,j}} = (d_{i,j} - \hat{d}_{i,j})\overrightarrow{u_{i,j}} \tag{3.10}$$

$$\overrightarrow{F_i} = \sum_{j \in H_i} \overrightarrow{F_{i,j}} \tag{3.11}$$

Each unknown node then updates its position in the direction of the resulting vector force in small increments over several iterations (with the force being recalculated at each step). However, it should be kept in mind that this technique may be susceptible to local minima.

3.6.9 Multi-dimensional scaling

Given a network with a sparse set of reference nodes, and a set of pair-wise distances between neighboring nodes (including reference and unknown nodes), another network localization approach utilizes a data analysis technique known as multi-dimensional scaling (MDS) [193]. It consists of the following three steps:

1. Use a distance-vector algorithm (similar to DV-distance) to generate an $n \times n$ matrix **M**, whose (i, j) entry contains the estimated distance between nodes i and j.

2. Apply classical metric-MDS to determine a map that gives the locations of all nodes in relative coordinates. The classical metric MDS algorithm is a matrix-based numerical technique that solves the following least squares problem: if the estimated distance matrix M can be expressed as the sum of the actual distance matrix D and a residual error matrix E (i.e. $M = D + E$), then determine possible locations for all n nodes, such that the sum of squares of the elements of E is minimized.

3. Take the position of reference nodes into account to obtain normalized absolute coordinates.

3.6.10 Reference-less localization

In some scenarios, we may encounter sensor networks that are deployed in such an *ad hoc* manner, without GPS capabilities, that there are no reference nodes whatsoever. In such a case, the best that can be hoped for is to obtain the location of the network nodes in terms of relative, instead of absolute, coordinates. While such a map is not useful for location stamping of sensor data, it can be quite useful for other functions, such as providing the information required to implement geographic routing schemes.

The multi-dimensional scaling problem (described above) can provide such a relative map, by simply eliminating step 3. Rao *et al.* [170] also develop such a technique for creating a virtual coordinate system for a network where there are no reference nodes and also where no distance estimates are available (unlike with MDS). Their algorithm is described as a progression of three scenarios with successively fewer assumptions:

1. All (and only) nodes at the boundary of the network are reference nodes.
2. Nodes at the boundary are aware that they are at the boundary, but are not reference nodes.
3. There are no reference nodes in the network, and no nodes are aware that they are at the boundary.

In the first scenario, all nodes execute a simple iterative algorithm for localization. Unknown interior nodes begin by assuming a common initial coordinate (say [0,0]), then at each step, each unknown node determines its location as the centroid of the locations of all its neighbors. It is shown that this algorithm tends to "stretch" the locations of network nodes through the location region. When the algorithm converges, nodes have determined a location that is close to their nearest boundary nodes. Figure 3.6 gives an example of a final solution.

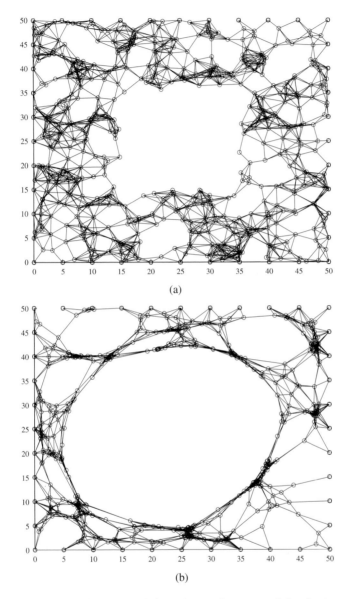

(a)

(b)

Figure 3.6 An illustration of the reference-less network localization technique assuming boundary node locations are known: (a) original map and (b) obtained relative map

   While the final solution is generally not accurate, it is shown that for greedy geographic routing it results in only slightly longer routing paths and potentially even slightly better routing success rates (as non-ideal positions can sometimes improve over the local optima that arise in greedy geographic routing).

The second scenario can be reduced approximately to the first. This can be done by having the border nodes first flood messages to communicate with each other and determine the pair-wise hop-counts between themselves. These hop-counts are then used in a triangulation algorithm to obtain virtual coordinates for the set B of all border nodes by minimizing

$$\sum_{i,j \in B} (\text{hops}(i, j) - \text{dist}(i, j))^2 \qquad (3.12)$$

where $\text{hops}(i, j)$ is the number of hops between border nodes i, j, and $\text{dist}(i, j)$ their Euclidean distance for given virtual coordinates. An additional bootstrapping mechanism ensures that all nodes calculate consistent virtual coordinates.

Finally, the third scenario can be reduced to the second. Any node that is farthest away from a common node in terms of hop-count with respect to all its two-hop neighbors can determine that it is on the border. This hop-count determination is performed through a flood from one of the bootstrap nodes.

3.7 Theoretical analysis of localization techniques

3.7.1 Cramér–Rao lower bound

One theoretical tool of utility in analyzing limitations on the performance of localization techniques is the use of the Cramér–Rao bound. The Cramér–Rao bound (CRB) is a well-known lower bound on the error variance of any unbiased estimator, and is defined as the inverse of the Fisher information matrix (a measure of information content with respect to parameters). The CRB can be derived for different assumptions about the localization technique (e.g. TOA-based, RSS-based, proximity-based, node/network localization).

The CRB has been used to investigate error performance of K-level quantized RSSI-based localization [155]. A special case is $K = 2$, which corresponds to proximity information (whether the node is within range or not). The lower bound can be improved monotonically with K, with about 50% improvement if K is large compared with just using proximity alone. On the other hand, $K = 8$ (three bits of RSS quantization) suffices to give a lower bound that is very close to the best possible. It is also found that the MLE estimator, which is a biased estimator, provides location errors with variance close to that observed with the CRB.

CRB analysis has also been used to investigate the performance of network localization under different densities, and shown to give similar trends to the iterative/collaborate multilateration technique [184]. The CRB-based analysis suggests that localization accuracy improves with network density, with diminishing returns once each node has about 6–8 neighbors on average. It also suggests, somewhat surprisingly, that increasing the fraction of beacon nodes from 4% to 20% does not dramatically decrease the localization error (under the assumptions of uniform placement, high-density, low-ranging error).

3.7.2 Unique network localization

There is a strong connection between the problem of unique network localization and a mathematical subject known as rigidity theory [46].

Definition 3
Consider a sensor network with n nodes (m reference nodes and $n - m$ unknown nodes) located in the 2D plane, with edges between neighboring nodes. Information is available about the exact location coordinates of the reference nodes, and the exact Euclidean distance between all neighboring nodes. This network is said to be **uniquely localizable** *if there exists only one possible assignment of (x, y) coordinates to all unknown nodes that is consistent with all the available information about distances and positions.*

The key result concerning the conditions for a network to be unique localizable is the following:

Theorem 6
A network N is uniquely localizable if and only if the weighted grounded graph G'_N corresponding to it is globally rigid.

There are two terms here that need to be explained – weighted grounded graph and global rigidity. The *weighted grounded graph* G'_N is constructed from the graph described by network N (with each edge weighed by the corresponding distance) by adding additional edges between all pairs of reference nodes, labelled with the distance between them (which can be readily calculated, since reference positions are known).

We shall give an intuitive definition of global rigidity. Consider a configuration graph of points in general position on the plane, with edges connecting some of them to represent distance constraints. Is there another configuration consisting of different points on the plane that preserves all the distance constraints on the edges (excluding trivial changes, such as translations, rotations, and mirror images)? If there is not, the configuration graph is said to be globally rigid in the plane. Figure 3.7 gives examples of non-globally rigid and globally rigid configuration graphs.

There exist polynomial algorithms to determine whether a given configuration graph is globally rigid in the plane, and hence to determine if a given network is uniquely localizable. However, the problem of realizing globally rigid weighted graphs (which is closely related to actually determining possible locations of

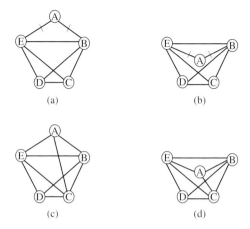

(a)   (b)

(c)   (d)

Figure 3.7 Examples of configuration graphs that are not globally rigid ((a),(b)) and that are globally rigid ((c),(d))

the unknown nodes in the corresponding network) is NP-hard. While this means that in the worst case there exist no known tractable algorithms to solve all instances, in the case of geometric random graphs, with at least three reference nodes within range of each other, there exists a critical radius threshold that is $O\left(\sqrt{\frac{\log n}{n}}\right)$, beyond which the network is uniquely localizable in polynomial time with high probability.

3.8 Summary

Determining the geographic location of nodes in a sensor network is essential for many aspects of system operation: data stamping, tracking, signal processing, querying, topology control, clustering, and routing. It is important to develop algorithms for scenarios in which only some nodes have known locations.

The design space of localization algorithms is quite large. The selection of a suitable algorithm for a given application and its performance depends upon several key factors, such as: what information about known locations is already available, whether the problem is to locate a cooperative node, how dynamic location changes are, the desired accuracy, and the constraints placed on hardware. On the basis of what needs to be localized, the location algorithms that have been proposed can be broadly classified into two categories: (i) node localization algorithms, which provide the location of a single unknown node given a number of reference nodes, and (ii) network localization algorithms, which provide the location of multiple unknown nodes in a network given other reference nodes. The node localization algorithms are often a building-block component of network localization algorithms. The accuracy of the localization algorithms is often dependent crucially upon how detailed the information obtained from reference nodes is.

The node localization algorithms we discussed include centroids, the use of overlapping geometric constraints, triangulation using distance estimates obtained using received signal strength and time difference of arrival, as well as AoA and pattern-matching approaches. For triangulation, TDoA techniques provide very accurate ranging at the expense of slightly more complex hardware. For RSS-based systems, an alternative to ranging-based triangulation for dense deployments is the ecolocation technique, which uses sequence-based decoding instead of absolute RSS values.

Network localization techniques include joint estimation techniques, iterative and collaborative multilateration, force-calculation, and multi-dimensional scaling. Even when no reference points are available, it is possible to construct a useful map of relative locations.

On the theoretical front, the Cramér–Rao bound on the error variance of unbiased estimators is useful in analyzing the performance of localization techniques. Rigidity theory is useful in formalizing the necessary and sufficient conditions for the existence of unique network localization.

Exercises

3.1 *Centroid:* Consider a node with unknown coordinates located in a square region of side d, with four reference nodes on the corners of the square. Consider three cases, when the reference beacons can be received within circles of radius: (a) $R = \frac{d}{\sqrt{2}}$, (b) $R = d$, and (c) $R = \sqrt{2}d$. In each case, identify the different unique centroid solutions that can be obtained (depending on the location of the unknown node) and the corresponding distinct regions. Estimate the worst case and average location estimate error in each case.

3.2 *Centroid versus proximity:* Consider three polygonal regions with reference nodes located at the vertices: (a) an equilateral triangle, (b) a square, and (c) a regular pentagon. Assume that an unknown node is uniformly likely to be located anywhere within the region, and that it is always within range of all reference nodes (and can correctly determine its nearest reference node if needed for a proximity determination). For each case compare the centroid localization technique with proximity localization, and indicate which performs better on average. Do you observe a trend in the three cases?

3.3 *Overlapping geometric constraints:* As in exercise 3.1, consider a node located in the square region of side d with four reference nodes A, B, C, and D located at the corners. Assuming $R = d$, identify the overlapping geometric region that the unknown node is constrained to be in if it can hear beacons from the three reference nodes A, B, C. Can you constrain the node's location further by taking explicitly into account the "negative information" that it does not hear from node D?

3.4 *ID-CODE:* Determine a minimum cardinality identifying code for the deployment region graph shown in Figure 3.8, and indicate the corresponding unique location identifiers for each region.

3.5 *RSSI-based distance estimates:* Using the statistical model of equation (3.2), assuming $d_0 = 1\text{m}, \eta = 3, \sigma = 7$, generate a scatter plot of RSSI-based estimated distance versus true distance. What do you observe?

3.6 *Distance-based position estimation:* Say there are three known reference nodes located at $(-2,0)$, $(1,1)$, $(2,-2)$ respectively. Use a least squares minimization approach to estimate the position of an unknown node, if the measured distances from these reference nodes to that unknown node are (a) 1.8, 1.2, and 3 respectively; and if they are (b) 2, $\sqrt{2}$, and $\sqrt{8}$ respectively. What is the resultant location error in each case?

3.7 *Sequence-based localization:* Consider an unknown node in a square region with reference nodes at the four corners, labelled clockwise from the top left as A, B, C, and D respectively.

   (a) What region must the unknown node lie in if there are no errors in the received sequence C, B, D, A?
   (b) What is the "best fit" location of the unknown node if the erroneous sequence A, B, C, D is received?

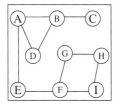

Figure 3.8 Graph for exercise 3.4

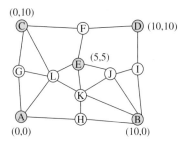

Figure 3.9 A network localization problem for exercise 3.8. Shaded nodes are reference nodes with known locations in parenthesis, all other nodes are unknown

3.8 *Iterative network localization:* For the network of Figure 3.9, determine the locations of all unknown nodes in a sequential manner using centroids. Specifically, calculate each unknown node's location in an appropriate sequence as the centroid of three neighboring known locations (either references or previously calculated locations).

3.9 *Reference-less localization:* Consider a square grid of 100 nodes located at coordinate points $(m/10, n/10)$, where m, n are each integers varying from 0 to 10. Assume that all the locations of border nodes are known *a priori* and that all other nodes are unknown location nodes that can communicate with their four immediate neighbors in the cardinal directions. Use the reference-less localization technique to determine the location of the unknown nodes iteratively. How does the final solution obtained in this case compare with the correct solution? In general, what are the conditions under which this technique can yield a relative location map that is very different from the correct location map?

3.10 *Graph rigidity:* Consider the graph represented by a regular polygon with n vertices and sides of distance d located in the 2-D plane.

(a) Prove that it is not globally rigid for $n > 3$.
(b) Derive an expression for the minimum number of additional edges that must be added to the graph to make it globally rigid, as a function of n.

4

Time synchronization

4.1 Overview

Given the need to coordinate the communication, computation, sensing, and actuation of distributed nodes, and the spatio-temporal nature of the monitored phenomena, it is no surprise that an accurate and consistent sense of time is essential in sensor networks. In this chapter, we shall discuss the many motivations for time synchronization, the challenges involved, as well as some of the solutions that have been proposed.

Distributed wireless sensor networks need time synchronization for a number of good reasons, some of which are described below:

1. **For time-stamping measurements:** Even the simplest data collection applications of sensor networks often require that sensor readings from different sensor nodes be provided with time stamps in addition to location information. This is particularly true whenever there may be a significant and unpredictable delay between when the measurement is taken at each source and when it is delivered to the sink/base station.
2. **For in-network signal processing:** Time stamps are needed to determine which information from different sources can be fused/aggregated within the network. Many collaborative signal processing algorithms, such as those for tracking unknown phenomena or targets, are coherent and require consistent and accurate synchronization.
3. **For localization:** Time-of-flight and TDoA-based ranging techniques used in node localization require good time synchronization.
4. **For cooperative communication:** Some physical layer multi-node cooperative communication techniques involve multiple transmitters transmitting

in-phase signals to a given receiver. Such techniques [105] have the potential to provide significant energy savings and robustness, but require tight synchronization.

5. **For medium-access:** TDMA-based medium-access schemes also require that nodes be synchronized so that they can be assigned distinct slots for collision-free communication.

6. **For sleep scheduling:** As we shall see in the following chapters, one of the most significant sources of energy savings is turning the radios of sensor devices off when they are not active. However, synchronization is needed to coordinate the sleep schedules of neighboring devices, so that they can communicate with each other efficiently.

7. **For coordinated actuation:** Advanced applications in which the network includes distributed actuators in addition to sensing require synchronization in order to coordinate the actuators through distributed control algorithms.

4.2 Key issues

The clock at each node consists of timer circuitry, often based on quartz crystal oscillators. The clock is incremented after each K ticks/interrupts of the timer. Practical timer circuits, particularly in low-end devices, are unstable and error prone. A model for clock non-ideality can be derived using the following expressions [89]. Let f_0 be the ideal frequency, Δf the frequency offset, d_f the drift in the frequency, and $r_f(t)$ an additional random error process, then the instantaneous oscillator frequency $f_i(t)$ of oscillator i at time t can be modeled as follows:

$$f_i(t) = f_0 + \Delta f + d_f t + r_f(t) \tag{4.1}$$

Then, assuming $t = 0$ as the initial reference time, the associated clock reads time $C_i(t)$ at time t, which is given as:

$$C_i(t) = C_i(0) + \frac{1}{f_0} \int_0^t f_i(\tau) d\tau$$

$$= C_i(0) + t + \frac{\Delta f}{f_0} t + \frac{d_f t^2}{2} + r_c(t) \tag{4.2}$$

where $r_c(t)$ is the random clock error term corresponding to the error term $r_f(t)$ in the expression for oscillator frequency. Frequency drift and the random error term may be neglected to derive a simpler linear model for clock non-ideality:

$$C_i(t) = \alpha_i + \beta_i t \qquad (4.3)$$

where α_i is the clock offset at the reference time $t = 0$ and β_i the clock drift (rate of change with respect to the ideal clock). The more stable and accurate the clock, the closer α_i is to 0, and the closer β_i is to 1. A clock is said to be *fast* if β_i is greater than 1, and *slow* otherwise.

Manufactured clocks are often specified with a maximum drift rate parameter ρ, such that $1 - \rho \leq \beta_i \leq 1 + \rho$. Motes, typical sensor nodes, have ρ values on the order of 40 ppm (parts per million), which corresponds to a drift rate of $\pm 40 \mu s$ per second.

Note that any two clocks that are synchronized once may drift from each other at a rate at most 2ρ. Hence, to keep their relative offset bounded by δ seconds at all times, the interval τ_{sync} corresponding to successive synchronization events between these clocks must be kept bounded: $\tau_{\text{sync}} \leq \delta/2\rho$.

Perhaps the simplest approach to time synchronization in a distributed system is through periodic broadcasts of a consistent global clock. In the US, the National Institute for Standards and Technology runs the radio stations WWV, WWVH, WWVB, that continuously broadcast timing signals based on atomic clocks. For instance WWVB, located at Fort Collins, Colorado, broadcasts timing signals on a 60 kHz carrier wave on a high power (50 kW) signal. Although the transmitter has an accuracy of about 1 μs, due to communication delays, synchronization around only 10μs is possible at receivers with this approach. While this can be implemented relatively inexpensively, the accuracy may not be sufficient for all purposes. Satellite-based GPS receivers can provide much better accuracy, of the order of 1μs or less, albeit at a higher expense, and they operate only in unobstructed environments. In some deployments it may be possible to use beacons from a subset of GPS-equipped nodes to provide synchronization to all nodes. In yet other networks, there may be no external sources of synchronization.

The requirements for time synchronization can vary greatly from application to application. In some cases the requirements may be very stringent – say 1μs synchronization between clocks – in others, it may be very lax – of the order of several milliseconds or even more. In some applications it will be necessary to keep all nodes synchronized globally to an external reference, while in others it will be sufficient to keep nodes synchronized locally and pair-wise to their immediate neighbors. In some applications it may be necessary to keep nodes synchronized at all times, and in other cases it may suffice only to know, *post facto*, the times when particular events occurred. Additional factors that determine the suitability of a particular synchronization approach to a given sensor network context include the corresponding energy costs, convergence times, and equipment costs.

4.3 Traditional approaches

Time synchronization is a long-studied subject in distributed systems, and a number of well-known algorithms have been developed for different conditions.[1]

For example, there is a well-known algorithm by Lamport [112] that provides a consistent ordering of all events in a distributed system, labelling each event x with a distinct time stamp L_x, such that: (a) $L_x \neq L_y$ for all unique events x and y, (b) if event x precedes event y within a node $L_x < L_y$, and (c) if x is the transmission of a message and y its reception at another node, $L_x < L_y$. These Lamport time stamps do not provide true causality. Say the true time of event x is indicated as T_x; then, while it is true that $T_x < T_y \rightarrow L_x < L_y$, it is not true that $L_x < L_y \rightarrow T_x < T_y$. Such a true causal ordering requires other approaches, such as the use of vector time stamps [209].

A fundamental technique for two-node clock synchronization is known as Cristian's algorithm [36]. A node A sends a request to node B (which has the reference clock) and receives back the value of B's clock, T_B. Node A records locally both the transmission time T_1 and the reception time T_2. This is illustrated in Figure 4.1.

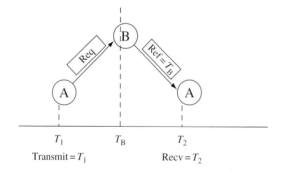

Figure 4.1 Cristian's synchronization algorithm

---

[1] The text by Tanenbaum and van Steen [209] provides a good discussion of many of these algorithms; we shall summarize these only briefly here.

In Cristian's time-synchronization algorithm, there are many sources of uncertainty and delay, which impact its accuracy. In general, message latency can be decomposed into four components, each of which contributes to uncertainty [108]:

- Send time – which includes any processing time and time taken to assemble and move the message to the link layer.
- Access time – which includes random delays while the message is buffered at the link layer due to contention and collisions.
- Propagation time – which is the time taken for point-to-point message travel. While negligible for a single link, this may be a dominant term over multiple hops if there is network congestion.
- Receive time – which is the time taken to process the message and record its arrival.

Good estimates of the message latency as well as the processing latency within the reference node must be obtained for Cristian's algorithm. The simplest estimate is to approximate the message propagation time as $(T_2 - T_1)/2$. If the processing delay is known to be I, then a better estimate is $(T_2 - T_1 - I)/2$. More sophisticated approaches take several round-trip delay samples and use minimum or mean delays after outlier removal.

The network time protocol (NTP) [138] is used widely on the Internet for time synchronization. It uses a hierarchy of reference time servers providing synchronization to querying clients, essentially using Cristian's algorithm.

4.4 Fine-grained clock synchronization

Several algorithms have been proposed for time synchronization in WSN. These all utilize time measurement-based message exchanges between nodes from time to time in order to synchronize the clocks on different nodes.

4.4.1 Reference broadcast synchronization (RBS)

The reference broadcast synchronization algorithm (RBS) [43] exploits the broadcast nature of wireless channels. RBS works as follows. Consider the scenario shown in Figure 4.2, with three nodes A, B, and C within the same broadcast domain. If B is a beacon node, it broadcasts the reference signal (which contains no timing information) that is received by both A and C simultaneously (neglecting propagation delay). The two receivers record the local time when the reference signal was received. Nodes A and C then exchange this local time stamp through separate messages. This is sufficient for the two receivers

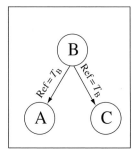 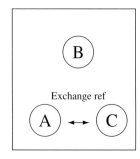

Figure 4.2 The reference broadcast synchronization (RBS) technique

to determine their relative offsets at the time of reference message reception. This basic scheme, which is quite similar to the CesiumSpray mechanism for synchronizing GPS-equipped nodes [213], can be extended to greater numbers of receivers. Improvements can be made by incorporating multiple reference broadcasts, which can help mitigate reception errors, as well as by estimating clock drifts.

A key feature of RBS is that it eliminates sender-side uncertainty completely. In scenarios where sender delays could be significant (particularly when time stamping has to be performed at the application layer instead of the link layer) this results in improved synchronization.

RBS can be extended to a multi-hop scenario as follows. If there are several separate reference beacons, each has its own broadcast domain that may overlap with the others. Receivers that lie in the overlapping region (in the broadcast domains of multiple references) provide "bridges" that allow nodes across these domains to determine the relationship between their local clocks, e.g., if nodes A and C are in range of reference B and nodes C and D are in range of reference E, then node C provides this bridge. In a large network, different paths through the temporal graph, representing connections between nodes sharing the same reference broadcast domain, provide different ways to convert times between arbitrary nodes. For efficiency, instead of computing these conversions *a priori* using global network information, these conversions can also be performed locally, on-the-fly, as packets traverse the network.

An interesting extension to the RBS technique is proposed and examined analytically in [103]. In particular, it is noted that the basic RBS scheme is composed only of a series of independent pair-wise synchronizations. This does not ensure global consistency in the following sense. Consider three nodes A, B, and C in the same domain, whose pair-wise offsets are determined through RBS.

There is no guarantee that the estimates obtained of $C_A(t) - C_B(t)$ and $C_B(t) - C_C(t)$ add up to the estimate obtained for $C_A(t) - C_C(t)$. An alternative technique has been developed for obtaining globally consistent minimum-variance pair-wise synchronization estimates, based on flow techniques for resistive networks [103].

4.4.2 Pair-wise sender–receiver synchronization (TPSN)

The timing-sync protocol for sensor networks (TPSN) [55] provides for classical sender–receiver synchronization, similar to Cristian's algorithm. As shown in Figure 4.3, node A transmits a message that is stamped locally at node A as T_1. This is received at node B, which stamps the reception time as its local time T_2. Node B then sends the packet back to node A, marking the transmission time locally at B as T_3. This is finally received at node A, which marks the reception time as T_4.

Let the clock offset between nodes A and B be Δ and the propagation delay between them be d. Then

$$T_2 = T_1 + \Delta + d \tag{4.4}$$

$$T_4 = T_3 - \Delta + d \tag{4.5}$$

which then result in the following:

$$\Delta = \frac{(T_2 - T_4) - (T_1 - T_3)}{2} \tag{4.6}$$

$$d = \frac{(T_2 + T_4) - (T_1 + T_3)}{2} \tag{4.7}$$

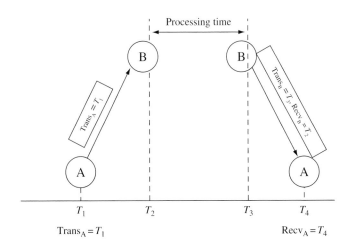

Figure 4.3 The basic sender–receiver synchronization technique used in TPSN

Network-wide time synchronization in TPSN is obtained level-by-level on a tree structure. Nodes at level 1 first synchronize with the root node. Nodes at level 2 then each synchronize with one node at level 1 and so on until all nodes in the network are synchronized with respect to the root.

Under the assumption that sender-side uncertainty can be mitigated by performing time stamping close to transmissions and receptions at the link layer, it is shown in [55] that the synchronization error with the pair-wise sender–receiver technique can actually provide twice the accuracy of the receiver–receiver technique used in RBS over a single link. This can potentially translate to even more significant gains over multiple hops.

The lightweight time synchronization (LTS) technique [68] is also a similar tree-based pair-wise sender–receiver synchronization technique.

4.4.3 Linear parameter-based synchronization

Another interesting approach to sender–receiver synchronization is the use of a linear model for clock behavior and corresponding parameter constraints to obtain synchronization [198]. This approach aims to provide deterministic bounds on relative clock drift and offset. Recall the simplified linear relationship for the behavior of a clock in equation (4.3). By combining that equation for nodes A and B together, we can derive the relationship between any two clocks A and B:

$$C_A(t) = \alpha_A - \alpha_B \frac{\beta_A}{\beta_B} + \frac{\beta_A}{\beta_B} C_B(t) \tag{4.8}$$

which can be re-written simply as

$$C_A(t) = \alpha_{AB} + \beta_{AB} C_B(t) \tag{4.9}$$

Assuming the same pair-wise message exchange as in TPSN for nodes A and B, we have that the transmission time T_1 and reception time T_4 are measured in node A's local clock, while reception time T_2 and transmission time T_3 are measured in node B's local clock. We therefore get the following temporal relationships:

$$T_1 < \alpha_{AB} + \beta_{AB} T_2 \tag{4.10}$$

$$T_4 > \alpha_{AB} + \beta_{AB} T_3 \tag{4.11}$$

The principle behind this approach to synchronization is to use these inequalities to determine constraints on the clock offset and drift. Each time such a pair-wise message takes place the expressions (4.10) and (4.11) provide additional constraints that together result in upper and lower bounds on the feasible

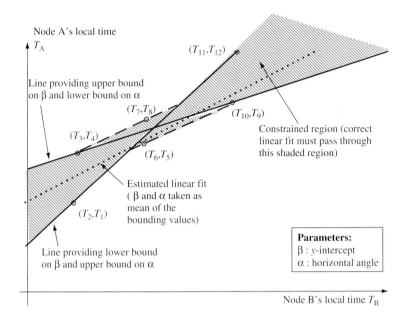

Figure 4.4 Visualization of bounding linear inequalities in the linear parameter-based technique

values of α_{AB} and β_{AB}. These bounds can be used to generate estimates of these quantities, which are needed for synchronization. The technique is illustrated in Figure 4.4.

To keep the storage and computational requirements of this approach bounded, there are two approaches. The first, called Tiny-Sync, keeps only four constraints at any time but can be suboptimal because it can discard potentially useful constraints. The second approach, called Mini-Sync, is guaranteed to provide optimal results and in theory may require a large number of constraints, but in practice is found to store no more than 40 constraints.

4.4.4 Flooding time synchronization protocol (FTSP)

The flooding time synchronization protocol (FTSP) [134] aims to further reduce the following sources of uncertainties, which exist in both RBS and TPSN:

1. **Interrupt handling time:** This is the delay in waiting for the processor to complete its current instruction before transferring the message in parts to the radio.

2. **Modulation/encoding time:** This is the time taken by the radio to perform modulation and encoding at the transmitter, and the corresponding demodulation and decoding at the receiver.

FTSP uses a broadcast from a single sender to synchronize multiple receivers. However, unlike RBS, the sender actually broadcasts a time measurement, and the receivers do not exchange messages among themselves. Each broadcast provides a synchronization point (a global–local time pair) to each receiver. FTSP has two main components:

1. Multiple time measurements: The sender takes several time stamp measurements during transmission, one at each byte boundary after a set of SYNC bytes used for byte alignment. These measurements are normalized by subtracting an appropriate multiple of the byte transmission time, and only the minimum of these multiple measurements is embedded into the message. At the receiver too, multiple time measurements are taken and the minimum of those is used as the receiver time. This serves to reduce the jitter significantly in interrupt handling and the (de)coding and (de)modulation times. With as few as six time stamps, an order of magnitude improvement in precision can be obtained on a Mica Mote platform (from the order of tens of microseconds to the order of about one microsecond).
2. Flooded messaging: To propagate the synchronization information, a flooding approach is used. First, a single uniquely identifiable node in the network provides the global clock. The reception of each broadcast message allows a receiver to accumulate a reference synchronization point. When a receiver accumulates several reference points, it becomes synchronized itself (e.g. using a regression line to estimate the local clock drift). Nodes can collect reference points either from the global reference node, or from other nodes that are already synchronized. The frequency of the flooding provides a tradeoff between synchronization accuracy and overhead.

4.4.5 Predictive time synchronization

The simple linear model of equation (4.3) is only reasonable for very short time intervals. In the real world, clock drift can vary over time quite drastically due to environmental temperature and humidity changes. This is the reason clock drifts must be continually reassessed. The naïve approach to this reassessment is to re-synchronize nodes periodically at the same interval. However, a static synchronization period must be chosen conservatively to accommodate a range of environments. This does not take into account the possibility of temporal correlations in clock drift. It will thus incur an unnecessarily high overhead in many cases.

This problem is addressed by the predictive synchronization mechanism [56] in which the frequency of inter-node time sampling is adaptively adjusted. It has been determined through an empirical study that environments are characterized by a time constant T over which drift rates are highly correlated, which can be determined through a learning phase. Depending on the time sampling period S, a window of T/S prior sample measurements is used in this technique not only to predict the clock drift (through linear regression), but also to estimate the error in the prediction. A MIMD technique is used to adapt the sampling period: if the prediction error is above a desirable threshold, the sampling period S is reduced multiplicatively; and if it is below threshold, the sampling period is increased accordingly. This adaptive scheme provides for robust long-term synchronization in a self-configuring manner.

4.5 Coarse-grained data synchronization

The Wisden system (a wireless sensor network system for structural-response data acquisition [231]) presents an excellent lightweight alternative to clock-synchronization approaches that is suitable for data-gathering applications. The approach is to collect and record latency measurements within each packet in a special residence time field as the packet propagates through the network. The estimate of delivery latency recorded in the packet is then used to provide a retroactive time stamp eventually at the base station when the packet arrives.

In this approach, only the base station is required to have an accurate reference clock. Since the radio propagation delays are insignificant, what is measured at each hop is actually the time that the packet spends at each node – which can be of the order of milliseconds due to queuing and processing delays. Say the time spent by packet i at the kth intermediate node on an $n+1$ hop-path to the destination is λ_k^i, let the time at the base station d when the packet is received be T_d^i, then the packet's original contents are time stamped to have been generated at source s at

$$T_s^i = T_d^i - \sum_{k=1}^{n} \lambda_k^i \tag{4.12}$$

This approach assumes that time stamps can be added as close to the packet transmission and reception as possible at the link layer. It is thus robust to many of the sources of latency uncertainty that contribute to error in other synchronization approaches. However, it is vulnerable to varying clock drifts at the intermediate nodes, particularly when packet residence times within the network are long. For example, it has been estimated that with a 10 ppm drift, data samples cannot reside for more than 15 minutes before accumulating 10 ms of error in the time stamp.

4.6 Summary

Like localization, time synchronization is also a core configuration problem in WSNs. It is a fundamental service building block useful for many network functions, including time stamping of sensor measurements, coherent distributed signal processing, cooperative communication, medium-access, and sleep scheduling. Synchronization is necessitated by the random clock drifts that vary depending on hardware and environmental conditions.

   Two approaches to fine-grained time synchronization are the receiver–receiver synchronization technique of RBS, and the more traditional sender–receiver approach of TPSN. While the latter provides for greater accuracy on a single link, RBS has the advantage that multiple receivers can be synchronized with fewer messages. It has been shown that these can provide synchronization of the order of tens of micro-seconds. The flooding-time synchronization protocol further improves performance by another order of magnitude by reducing uncertainties due to jitter in interrupt handling and coding/modulation. Thus it appears that even fairly demanding synchronization requirements can be met in principle through such algorithms. However, there is an energy–accuracy tradeoff involved in long-term synchronization, because the accuracy is determined by the frequency with which nodes are periodically re-synchronized. It has been shown that adaptive prediction-based drift-estimation techniques can reduce this overhead further.

   For some applications, instead of using inter-node synchronization, coarse-grained data time stamps can be obtained by timing packets as they move through the network and by performing a simple calculation at the final destination.

Exercises

4.1   *Unsynchronized clock drift:* Plot the variation of a clock's measured time with respect to real time, assuming $f_0 = 1$, $\Delta f = 0.1$, $d_f = -0.01$.

4.2   *Synchronization frequency:* How frequently should a clock with $\Delta f = 0.3$, $d_f = 0$ be synchronized with an ideal clock to ensure that it does not stray more than one time unit from the real clock?

4.3 *Communication scaling:* Assume there are n nodes within range of each other, among which one has an ideal clock. How many messages in total are required to synchronize the clocks of all nodes in case of (a) RBS and (b) TPSN? What does this suggest about the scalability of these techniques?

4.4 *Error propagation:* Consider n nodes in a chain, numbered 1 to n left to right. Assume that, starting with node 2, each node's clock is successively synchronized pair-wise with the node on its left. Due to processing-time uncertainties, assume that the clock difference between each pair of neighboring nodes after synchronization (i.e. the synchronization error) is not deterministically zero, but rather a zero-mean Gaussian with variance σ^2. What is the form of the end-to-end synchronization error across n hops?

4.5 *Linear parameter-based synchronization:* Consider the following three sets of four time stamps on messages exchanged between the nodes A and B: [(1,2,3,8), (9,10,11,12), (13,18,19,20)]. As in Figure 4.4, draw a diagram showing the corresponding linear inequalities, indicating the constrained region as well as the lines corresponding to the bounds on α_{AB} and β_{AB}. What are these bounds and the corresponding mean estimate for synchronization?

4.6 *Data-stamping error accumulation:* Assuming a worst case drift of 40 ppm at each intermediate node, how long can a packet stay within the network when using the Wisden data-stamping technique before the total error exceeds 1 second?

Wireless characteristics

5.1 Overview

Wireless communication is both a blessing and a curse for sensor networks. On the one hand, it is key to their flexible and low-cost deployment. On the other hand, it imposes considerable challenges because wireless communication is expensive and wireless link conditions are often harsh and vary considerably in both space and time due to multi-path propagation effects.

Wireless communications have been studied in depth for several decades and entire books are devoted to the subject [171, 207]. The goal of this chapter is by no means to survey all that is known about wireless communications. Rather, we will focus on three sets of simple models that are useful in understanding and analyzing higher-layer networking protocols for WSN:

1. **Link quality model:** a realistic model showing how packet reception rate varies statistically with distance. This incorporates both an RF propagation model and a radio reception model.
2. **Energy model:** a realistic model for energy costs of radio transmissions, receptions, and idle listening.
3. **Interference model:** a realistic model that incorporates the capture effect whereby packets from high-power transmitters can be successfully received even in the presence of simultaneous traffic.

5.2 Wireless link quality

The following is the basic ideal model of a wireless link: two nodes have a perfect link (with 100% packet reception rate) if they are within communication range R, and a non-existent link (0% packet reception rate) if they are outside this range. This ideal model, as we have already seen in the preceding chapters, is the basis of many algorithms and is widely used in analytical and simulation studies. While it is useful in some contexts, the ideal model can be quite misleading when designing and evaluating routing protocols. Therefore we seek more realistic models based on real-world observations.

5.2.1 Empirical observations

Several researchers have undertaken experimental studies of link quality in WSN [57, 242, 225, 204]. The following are several key observations from these studies:

1. The packet reception contour formed by receptions at different locations from the same transmitter is not regular or isotropic (see Figure 5.1).
2. The link quality distributions with and without power control are highly dependent on environment and individual hardware differences. Indoor office environments, for instance, show worse link quality distributions than clutter-free outdoor settings. Changing the transmitter or receiver at the same location can change the link quality.
3. There are three distinct regions of link quality: a nearby connected region where packet reception rates are consistently high; beyond this lies a transitional region (also referred to as a gray area), where reception rates are highly variant (see Figure 5.2).
4. If a link has a very high packet reception rate (> 99%), then the link is firmly within the connected region and is likely to be quite reliable over time.
5. In the transitional region, there can be links that have excellent quality, although the node pairs are relatively far apart, and conversely there can be weak links that have poor quality, despite the relative proximity of the node pairs.
6. Particularly in the transitional region, there can be a significant number of asymmetric links, i.e. have a high link quality (high PRR) in one direction and low link quality in the other direction.
7. The nodes in the transitional region also see high time variation in the link quality.
8. The width of the transitional region can be quite significant as a fraction of the connected region. The width of this region depends very much on the operational environment.

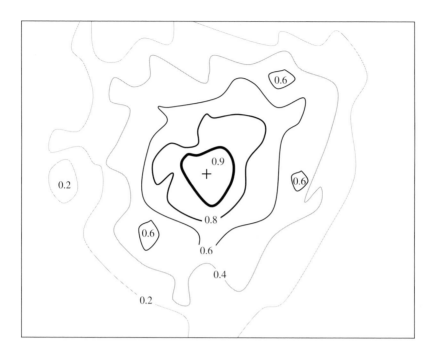

Figure 5.1 A realistic packet reception rate contour

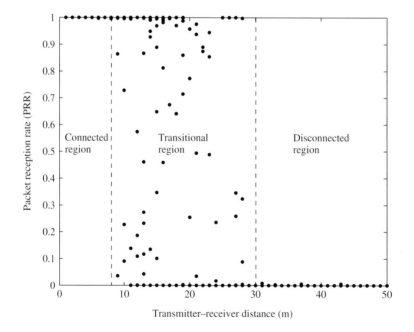

Figure 5.2 Realistic packet reception rate statistics with respect to inter-node distance

The main lesson to take away is that the transitional region is of particular concern in WSN as it contains high variance, unreliable links. As we shall see, the transitional region can be explained and understood using simple concepts from communication theory.

5.2.2 RF propagation model

While there are many sophisticated statistical models to describe multi-path fading in radio propagation (such as the Rayleigh, Ricean, and Nakagami distributions), several studies have shown that the following simple model showing a path-loss exponent and log-normal fading gives a very good match for short-range, line-of-sight communication, particularly indoor environments [149]:

$$P_{r,dB}(d) = P_{t,dB} - PL_{dB}(d) \tag{5.1}$$

$$PL_{dB}(d) = PL_{dB}(d_0) + 10\eta\log_{10}\left(\frac{d}{d_0}\right) + X_{\sigma,dB} \tag{5.2}$$

In the above expressions, $P_{r,dB}(d)$ is the received power in dB, $P_{t,dB}$ is the transmitted power, $PL_{dB}(d)$ is the path-loss in dB a distance d from the transmitter, d_0 is a reference distance, η is the path-loss exponent, which indicates the rate at which the mean signal decays with respect to distance, and $X_{\sigma,dB}$ is a zero-mean Gaussian random variable (in dB) with standard deviation σ. Figure 5.3 illustrates this model.

This basic model could be extended in many ways. Obstacles including walls can be modeled by adding an additional absorption term to the path-loss. The random fading term X_σ could be modeled as a multi-dimensional random process to incorporate both temporal and spatial correlations. Even richer models that explicitly characterize the impact of other factors besides distance – e.g. the antenna orientation and height – may be needed for some studies.

5.2.3 Radio reception model

The bit error rate for a given radio is a function of the received signal-to-noise ratio (SNR). The exact form of this function depends on the physical layer particulars of the radio, particularly the modulation and encoding scheme used. Depending on the frame size, and any frame-level encoding used, this can in turn be used to derive the relationship between the packet reception rate (PRR) and the receiver SNR. For instance, for a Mica 2 Mote, which uses a non-coherent

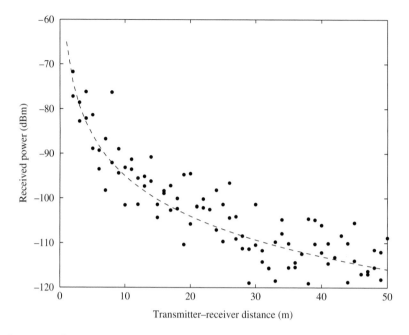

Figure 5.3 Illustration of received signal strength versus distance in a path-loss model with log-normal variance

FSK radio, the packet reception rate for a packet of length L bytes is given as the following function of the SNR [245]:

$$PRR = \left(1 - \frac{1}{2} \exp^{-\frac{SNR}{2}\frac{1}{0.64}} \right)^{8L} \tag{5.3}$$

Figure 5.4 shows how the PRR varies with the received signal strength, based on both analytical derivation and empirical measurements for a typical WSN node. The curve is sigmoidal and it is most important to note that there are two significant radio thresholds with respect to the received signal strength: a lower threshold below which the PRR is close to zero and a higher threshold beyond which it is close to one.

5.2.4 The transitional region

The composition of the received power versus distance curve with the upper and lower SNR thresholds for packet receptions yields the PRR versus distance behavior for links [245]. Figure 5.5 illustrates this composition, along with the three distinct regions observed empirically, with respect to distance: the connected

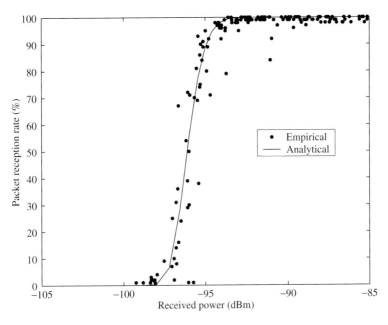

Figure 5.4 Experimental measurement of packet reception rates as a function of received radio signal strength in an indoor setting

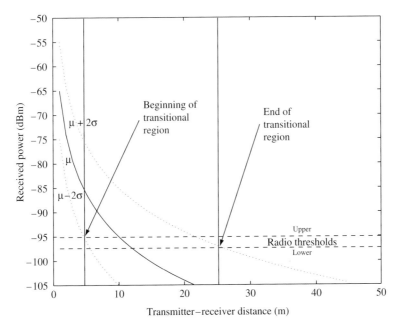

Figure 5.5 Composition of the RF propagation model and the radio reception model explains the empirical observations of three distinct regions

region, the transitional region, and the disconnected region. In the connected region, which occurs at close distances, the received signal strength is sufficient to provide an SNR higher than the upper threshold with high probability. In the disconnected region, which occurs at far distances, the signal strength is so low that the SNR is below the lower threshold with high probability. At distances in between, we have the transitional region, where there are significant link quality variations as the signal strength values cause the SNR to fluctuate between the two thresholds due to fading effects. This approach allows a nice decoupling of environmental parameters (η, σ) and radio parameters (lower and upper SNR thresholds). It can be used in simulations to generate link PRR statistics, as in Figure 5.2.

It should be noted that in the transitional region, even if there is no time variation due to lack of mobility, different links will show very different qualities. Minor receiver variations in the transitional region can cause high incidence of asymmetric links if the SNR fluctuates between the radio thresholds. And, if there is time-varying fading, its effects will be even more pronounced in the transitional region. Thus we can see that the transitional region is of particular concern, and has a big impact on the robustness of the wireless network.

This concern is further exacerbated by the fact that the area of the transitional region, particularly in comparison with the connected region, can be quite large. Even if the width of the transitional region (in distance) is the same as that of the connected region, because of the quadratic increase in area with respect to radius there will three times as much area covered by the transitional region. This, in turn, implies three times as many unreliable links as reliable links in the WSN.

Let us define the *transitional region (TR) coefficient* Γ, as the ratio of the radii of the transitional and connected regions. In view of the relative undesirability of the transitional region with respect to the connected region, generally it is better for the TR coefficient to be as low as possible.

It can be shown that the TR coefficient does not vary with transmission power, because both the connected and transitional regions grow proportionally with the transmission power. Table 5.1 shows how the coefficient behaves with respect to σ and η. It is understandable that with lower σ, when the variations due to fading are low, the relative size of the transitional region is smaller. The table also shows that, paradoxically, higher η environments characterized by rapid transmission power decay also show a lower TR coefficient, at the cost of a smaller connected region (though this can be combatted through power control).

One way to deal with the unreliability introduced by the transitional region in practice is to periodically monitor the quality of links and "blacklist" any links determined to be of poor quality (e.g. weak links with low reception rate,

Table 5.1 *The transitional region coefficient as a function of environmental parameters*

	$\sigma = 2$	$\sigma = 4$	$\sigma = 6$	$\sigma = 8$
$\eta = 2$	2.3	4.2	7.2	12.0
$\eta = 4$	0.8	1.3	1.9	2.6
$\eta = 6$	0.5	0.7	1.0	1.4
$\eta = 8$	0.3	0.5	0.7	0.9

asymmetric links) [225, 204, 62]. Properly implemented, blacklisting can provide a useful abstraction of an ideal topology for efficient communication and routing.

5.3 Radio energy considerations

Table 5.2 shows typical power consumption costs for a number of radio hardware devices for the different operating modes, as well as the time taken for turning the radio on to an active mode from sleep mode. The time to switch a radio back to sleep mode is not shown in this table, but it is generally negligible (on the order of $10\mu s$). While various radios differ in the absolute power values, depending on the output power and range, a common feature that should be observed in their power costs is that sleep mode generally consumes three orders of magnitude less power, while receive power and transmit power costs are of the same magnitude, with the receive–transmit power ratio ranging from about 1:1 to 1:2. The power costs for (a) keeping the mode in receive mode without actively receiving packets, (b) keeping the node in receive mode, while receiving packets, and for (c) keeping the node in receive mode while overhearing packets intended for others, are often very similar (within about 20–30%). These observations suggest that it is best to keep a radio in the sleep mode as much as possible (to cut down on idle receive mode costs) and for a long duration each time (to minimize switching costs).

The following is a simple model of energy cost per bit for communication over a link of distance d:

$$E = E_{tx} + E_{rx} = \alpha + \beta d^{\eta} \tag{5.4}$$

The distance-independent term α represents the energy cost of transmitter and receiver electronics, β represents a transmit amplifier constant, while d^{η}

Table 5.2 *Typical power consumption costs and startup times for different radios*

Radio	Frequency/Data Rate	Sleep	Receive	Transmit	Startup
CC 2420 [18]	2.4 GHz, 250 kbps	60 μW	59 mW	52 mW	0.6 ms
CC 1000 [17]	868 MHz, 19.2 kbps	0.6 μW	29 mW	50 mW	2.0 ms
MIT μAMPS-1 [139]	2.4 GHz, 1 Mbps	negligible	279 mW	330 mW	0.5 ms
IEEE 802.11b [165]	2.4 GHz, 11 Mbps	negligible	1.4 W	2.25 W	1.0 ms

(where η is the path-loss exponent) captures the amplification required to ensure constant power reception at the receiver. A striking observation made by Min and Chandrakasan [140] is that for most short-range radios $\alpha > \beta d^{\eta}$. Taking into account the fact that the distance-independent term can dominate the transmission cost, they conclude that, for practical short-range radios, energy costs may not be reduced significantly (if at all) by travelling multiple shorter hops with reduced output power.

5.4 The SINR capture model for interference

Recall the ideal circular communication range link model; a related ideal model is also often used in the analysis and design of MAC protocols. In this model, the interference caused by a node's transmissions is modeled to be the same as its ideal circular communication range (see Figure 5.6(a)). No other node within this range may receive a packet successfully due to collisions, while this node is transmitting. A variant of this model is one in which the disc of interference has a radius that is larger than for the disc of communication. While these models – in which good communication on a link implies that the transmissions on that link will always collide with other traffic incoming to the receiving node – offer very simple abstractions that are useful for design and analysis, they can also be potentially highly misleading.

In practice, radios are capable of receiving a packet error free, even when another packet is being transmitted by a neighbor (see Figure 5.6(b)). This is known as the *capture effect*. The following is a realistic model incorporating the capture effect.

Let $g_{i,j}$ be the channel gain on the link between nodes i and j (incorporating the mean path-loss as a function of distance as well as the log-normal fading component), P_i the output transmit power at node i, and N_i the noise power at

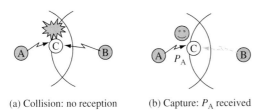

(a) Collision: no reception   (b) Capture: P_A received

Figure 5.6 Interference in wireless medium: (a) an idealized model, (b) the capture effect model. In the capture effect model simultaneous successful receptions are possible so long as SINR is sufficiently high at each receiver

node i. Then node 1 can receive a packet successfully from node 0 even if there is a set of interfering nodes I that are simultaneously transmitting packets if

$$\frac{P_0 g_{0,1}}{\sum_{i \in I} g_{i,1} P_i + N_1} > c \tag{5.5}$$

The expression on the left-hand side of this inequality is referred to as the signal-to-interference-plus-noise ratio (SINR), and the quantity c on the right-hand side is referred to as the *capture ratio* or capture threshold. The capture ratio c is dependent on the modulation and coding characteristics of the radio. When c is small, it is easier to capture a packet even in the presence of interfering transmissions.

This capture effect has been measured experimentally in sensor motes [226]. Through experiments on the CC1000-equipped Mica 2 platform, the authors of [226] found that up to 70% of all simultaneous in-communication-range transmissions exhibit the capture phenomenon. Compared with the idealized model, the capture phenomenon has the practical effect of reducing the levels of packet-loss in a contention-based MAC.

5.5 Summary

The wireless nature of the sensor networks we examine brings both advantages in terms of flexibility and cost as well as great challenges, because of the harsh radio channels. While radio propagation and fading models have been studied in depth for many years for mobile personal communications, their implication for multi-hop networks is only now beginning to be studied and understood.

A number of empirical studies have shown that designing and analyzing multi-hop wireless networks using high-level zero–one abstractions for links can be quite misleading. In terms of distance, three distinct regions of link quality have been identified: the connected region, where links are always high quality; the disconnected region, where links rarely exist; and the intermediate transitional region. The transitional region is of particular concern because it is considerable in size and contains many links that are dynamically varying and asymmetric. A simple realistic statistical model of packet reception rates with respect to transmitter–receiver distance can be derived and used to analyze the transitional region and to simulate realistic wireless topologies.

Communications can be a significant source of energy consumption in wireless networks, and the data suggest that it is important to minimize radio idle receive mode time by turning off the radio as much as possible when not in use; switching costs must also be kept in mind. The distance-independent term in the radio energy model can be quite significant, so that tuning the transmission power down for short-range transmissions may not provide large energy gains.

Depending on the modulation scheme used and the particular placement of communicating nodes, simultaneous transmissions within the same neighborhood can take place without significant packet loss. This effect is best modeled using the SINR-based capture model.

Exercises

5.1 *The transitional region:* Calculate the width of the transitional region, assuming the path-loss exponent $\eta = 3$, the log-normal fading variance $\sigma = 7$, the received power at a distance of 1 meter from a transmitter is -65 dBm, and the radio lower and upper thresholds for received signal strength are -105 dBm and -98 dBm respectively. What is the TR coefficient in this case?

5.2 *Neighborhood link quality:* For the scenario in exercise 5.1, if nodes are distributed uniformly with a high density, what percentage of a node's neighbors (defined here as nodes that are within the connected or transitional regions) will lie in the transitional region?

5.3 *Impact of radio characteristics on TR:* For the scenario in exercise 5.1, assume the radio was changed to one that is slightly worse, having lower and upper thresholds of -95 dBm and -90 dBm. What are the width of the transitional region and the value of the TR coefficient in this case? Is this better or worse?

5.4 *Energy model and hopping:* Assume the energy required to transmit a packet a distance of d is $\alpha + \beta d^{\eta}$. Let the optimization goal be to minimize the total total energy expenditure required to move information from a given source to a destination located a distance D away. Assume that nodes can be placed arbitrarily between these two nodes for the purpose of relaying. Derive an expression for the optimal number of relay hops. Comment on how the expression for the optimal number of hops varies with respect to η and α. Let $\alpha = 400$, $\beta = 0.6$, $\eta = 3$; what is the optimal number of hops to move information by a distance $D = 1000$?

5.5 *Capture with power control:* Consider two transmitters both a distance d from a receiver. The first transmitter has a transmitting power of P_1, the second transmits at a power of P_2. The received power at a distance d from a transmitter with power P_t is given as $P_t d^{-\eta}$. Assume the receiving node's radio has a capture threshold c and receiver noise power N. If $P_2 = -70\,\mathrm{dBm}$, $d = 20\,m$, $\eta = 3$, $c = 10\,\mathrm{dB}$, $N = -100\,\mathrm{dBm}$, what should P_1 be in order to ensure that the first transmitter's packets are received successfully at the receiver, even if both transmit simultaneously?

5.6 *Capture based on distance:* Consider a receiver and two transmitters both communicating with a common power P_t. The distance from the first transmitter to the receiver is d_1 and from the second transmitter to the receiver is d_2. If $P_t = -70\,\mathrm{dBm}$, $d_1 = 20\,m$, $\eta = 3$, $c = 10\,\mathrm{dB}$, $N = -100\,\mathrm{dBm}$, how far away should the second transmitter be from the receiver to ensure that the first transmitter's packets are received successfully, even if both transmit simultaneously?

6

Medium-access and sleep scheduling

6.1 Overview

An essential characteristic of wireless communication is that it provides an inherently shared medium. All medium-access control (MAC) protocols for wireless networks manage the usage of the radio interface to ensure efficient utilization of the shared bandwidth. MAC protocols designed for wireless sensor networks have an additional goal of managing radio activity to conserve energy. Thus, while traditional MAC protocols must balance throughput, delay, and fairness concerns, WSN MAC protocols place an emphasis on energy efficiency as well.

We shall discuss in this chapter a number of contention-based as well as schedule-based MAC protocols that have been proposed for WSN. A common theme through all these protocols is putting radios to a low-power "sleep mode" either periodically or whenever possible when a node is neither receiving nor transmitting.

6.2 Traditional MAC protocols

We begin with a focus on contention-based MAC protocols. Contention-based MAC protocols have an advantage over contention-free scheduled MAC protocols in low data rate scenarios, where they offer lower latency characteristics and better adaptation to rapid traffic variations.

6.2.1 Aloha and CSMA

The simplest forms of medium-access are unslotted Aloha and slotted Aloha. In unslotted Aloha, each node behaves independently and simply transmits a packet whenever it arrives; if a collision occurs, the packet is retransmitted after a random waiting period. The slotted version of Aloha works in a similar manner, but allows transmissions only in specified synchronized slots. Another classic MAC protocol is the carrier sense medium-access (CSMA) protocol. In CSMA, a node that wishes to transmit first listens to the channel to assess whether it is clear. If the channel is idle, the node proceeds to transmit. If the channel is busy, the node waits a random back-off period and tries again. CSMA with collision detection is the basic technique used in IEEE 802.3/Ethernet.

6.2.2 Hidden and exposed node problems

Traditional CSMA fails to avoid collisions and is inefficient in wireless networks because of two unique problems: the hidden node problem and the exposed node problems.

The hidden node problem is illustrated in Figure 6.1(a); here, node A is transmitting to node B. Node C, which is out of the radio range of A, will sense the channel to be idle and start packet transmission to node B too. In this case CSMA fails to avoid the collision because A and C are hidden to each other.

The exposed node problem is illustrated in Figure 6.1(b). Here, while node B is transmitting to node A, node C has a packet intended for node D. Because node C is in range of B, it senses the channel to be busy and is not able to send. However, in theory, because D is outside of the range of B, and A is outside of the range of C, these two transmissions would not collide with each other. The deferred transmission by C causes bandwidth wastage.

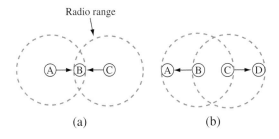

(a)   (b)

Figure 6.1 Problems with basic CSMA in wireless environments: (a) hidden node, (b) exposed node

These problems are duals of each other in a sense: in the hidden node problem packets collide because sending nodes do not know of another ongoing transmission, whereas in the exposed node problem there is a wasted opportunity to send a packet because of misleading knowledge of a non-interfering transmission. The key underlying mismatch is that it is not the transmitter that needs to sense the carrier, but the receiver. Some communication between the transmitter and receiver is needed to solve these problems.

6.2.3 Medium-access with collision avoidance (MACA)

The MACA Protocol by Karn [101] introduced the use of two control messages that can (in principle) solve the hidden and exposed node problems. The control messages are called *request to send* (RTS) and *clear to send* (CTS). The essence of the scheme is that when a node wishes to send a message, it issues an RTS packet to its intended recipient. If the recipient is able to receive the packet, it issues a CTS packet. When the sender receives the CTS, it begins to transmit the packet. When a nearby node hears an RTS addressed to another node, it inhibits its own transmission for a while, waiting for a CTS response. If a CTS is not heard, the node can begin its data transmission. If a CTS is received, regardless of whether or not an RTS is heard before, a node inhibits its own transmission for a sufficient time to allow the corresponding data communication to complete.

Under a number of idealized assumptions (e.g., ignoring the possibility of RTS/CTS collisions, assuming bidirectional communication, no packet losses, no capture effect) it can be seen that the MACA scheme can solve both the hidden node problem and the exposed node problem. Using the earlier examples, it solves the hidden node problem because node C would have heard the CTS message and suppressed its colliding transmission. Similarly it solves the exposed node problem because, although node C hears node B's RTS, it would not receive the CTS from node A and thus can transmit its packet after a sufficient wait.

6.2.4 IEEE 802.11 MAC

Closely related to MACA is the widely used IEEE 802.11 MAC standard [95]. The 802.11 device can be operated in infrastructure mode (single-hop connection to access points) or in *ad hoc* mode (multi-hop network). It also includes two mechanisms known as the distributed coordination function (DCF) and the point coordination function (PCF). The DCF is a CSMA-CA protocol (carrier sense multiple access with collision avoidance) with ACKs. A sender first checks to see if it should suppress transmission and back off because the medium is busy;

if the medium is not busy, it waits a period DIFS (distributed inter-frame spacing) before transmitting. The receiver of the message sends an ACK upon successful reception after a period SIFS (short inter-frame spacing). The RTS/CTS virtual carrier sensing mechanism from MACA is employed, but only for unicast packets. Nodes which overhear RTS/CTS messages record the duration of the entire corresponding DATA-ACK exchange in their NAV (network allocation vector) and defer access during this duration. An exponential backoff is used (a) when the medium is sensed busy, (b) after each retransmission (in case an ACK is not received), and (c) after a successful transmission.

In the second mechanism, PCF, a central access point coordinates medium-access by polling the other nodes for data periodically. It is particularly useful for real-time applications because it can be used to guarantee worst-case delay bounds.

6.2.5 IEEE 802.15.4 MAC

The IEEE 802.15.4 standard is designed for use in low-rate wireless personal area networks (LR-WPAN), including embedded sensing applications [94]. Most of its unique features are for a beacon-enabled mode in a star topology.

In the beacon-enabled mode for the star topology, the IEEE 802.15.4 MAC uses a superframe structure shown in Figure 6.2. A superframe is defined by a periodic beacon signal sent by the PAN coordinator. Within the superframe there is an active phase for communication between nodes and the PAN coordinator and an inactive phase, which can be adjusted depending on the sleep duty cycle desired. The active period has 16 slots that consist of three parts: the beacon, a contention access period (CAP), and a collision-free period (CFP) that allows for the allocation of guaranteed time slots (GTS). The presence of the collision-free period allows for reservation-based scheduled access. Nodes which communicate only on guaranteed time slots can remain asleep and need

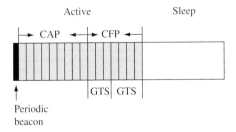

Figure 6.2 The superframe structure of IEEE 802.15.4 MAC

only wake-up just before their assigned GTS slots. The communication during CAP is a simple CSMA-CA algorithm, which allows for a small backoff period to reduce idle listening energy consumption. A performance evaluation of this protocol and its various settings and parameters can be found in [126].

While the IEEE 802.15.4 can, in theory, be used for other topologies, the beacon-enabled mode is not defined for them. In the rest of the chapter we will concern ourselves with both contention-based and schedule-based energy-efficient MAC protocols that are relevant to multi-hop wireless networks.

6.3 Energy efficiency in MAC protocols

Energy efficiency is obtained in MAC protocols essentially by turning off the radio to sleep mode whenever possible, to save on radio power consumption.

6.3.1 Power management in IEEE 802.11

There exist power management options in the infrastructure mode for 802.11. Nodes inform the access point (AP) when they wish to enter sleep mode so that any messages for them can be buffered at the AP. The nodes periodically wake-up to check for these buffered messages. Energy savings are thus provided at the expense of lower throughput and higher latency.

6.3.2 Power aware medium-access with signalling (PAMAS)

The PAMAS (power aware multi-access protocol with signalling) [199] is an extension of the MACA technique, where the RTS/CTS signalling is carried out on a separate radio channel from the data exchange. It is one of the first power aware MAC protocols proposed for multi-hop wireless networks. In PAMAS, nodes turn off the radio (go to sleep) whenever they can neither receive nor transmit successfully. Specifically they go to sleep whenever they overhear a neighbor transmitting to another node, or if they determine through the control channel RTS/CTS signaling that one of their neighbors is receiving. The duration of the sleep mode is set to the length of the ongoing transmissions indicated by the control signals received on the secondary channel. If a transmission is started while a node is in sleep mode, upon wake-up the node sends probe signals to determine the duration of the ongoing transmission and how long it can go back to sleep. In PAMAS, a node will only be put to sleep when it is inhibited from transmitting/receiving anyway, so that the delay/throughput performances of the network are not affected adversely. However, there can still be considerable energy wastage in the idle reception mode (i.e. the condition when a node has no packets to send and there is no activity on the channel).

6.3.3 Minimizing the idle reception energy costs

While PAMAS provides ways to save energy on overhearing, further energy savings are possible by reducing idle receptions. The key challenge is to allow receivers to sleep a majority of the time, while still ensuring that a node is awake and receiving when a packet intended for it is being transmitted. Based on the methods to solve this problem, there are essentially two classes of contention-based sensor network MAC protocols.

The first approach is completely asynchronous and relies solely on the use of an additional radio or periodic low-power listening techniques to ensure that the receiver is woken up for an incoming transmission intended for it. The second approach, with many variants, uses periodic duty-cycled sleep schedules for nodes. Most often the schedules are coordinated in such a way that transmitters know in advance when their intended receiver will be awake.

6.4 Asynchronous sleep techniques

In these techniques nodes normally keep their radios in sleep mode as a default, waking up briefly only to check for traffic or to send/receive messages.

6.4.1 Secondary wake-up radio

Nodes need to be able to sleep to save energy when they do not have any communication activity and be awake to participate in any necessary communications. The first solution is a hardware one – equipping each sensor node with two radios. In such a hardware design, the primary radio is the main data radio, which remains asleep by default. The secondary radio is a low-power wake-up radio that remains on at all times. Such an idea is described in the Pico Radio project [166], as well as by Shih *et al.* [195]. If the wake-up radio of a node receives a wake-up signal from another node, it responds by waking up the primary radio to begin receiving. This ensures that the primary radio is active only when the node has data to send or receive. The underlying assumption motivating such a design is that, since the wake-up radio need not do much sophisticated signal processing, it can be designed to be extremely low power. A tradeoff, however, is that all nodes in the broadcast domain of the transmitting node may be woken up.

6.4.2 Low-power listening/preamble sampling

El Hoiydi [83] and Hill and Culler [82] independently developed a similar rendezvous mechanism for waking up sleeping radios. In this technique, referred to
as preamble sampling or low-power listening, the receivers periodically wake-up
to sense the channel. If no activity is found, they go back to sleep. If a node
wishes to transmit, it sends a preamble signal prior to packet transmission. Upon
detecting such a preamble, the receiving node will change to a fully active receive
mode. The technique is illustrated in Figure 6.3.

The wake-up signal could potentially be sent over a high-level packet interface;
however, a more efficient approach is to implement this directly in the physical
layer – thus the wake-up signal may be no more than a long RF pulse. The
detecting node then only checks for the radio energy on the channel to determine
whether the signal is present. Hill and Culler argue that this can reduce the
receiver check duty cycle to as low as 0.00125%, allowing an almost 2000-fold
improvement in lifetime compared with a packet-level wake-up (from about a
week to about 38 years). We should note that this scheme will also potentially
wake-up all possible receivers in a given transmitter's neighborhood, though
mechanisms such as information in the header can be used to put them back to
sleep if the communication is not intended for them.

6.4.3 WiseMAC

Low-power listening/preamble sampling has one potential shortcoming: the long
preamble that the transmitter needs to send can in some situations cause throughput reduction and energy wastage for both sender and receiver. The WiseMAC
protocol [84] builds upon preamble sampling to correct this deficiency. Using

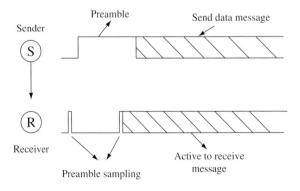

Figure 6.3 The low-power listening technique of preamble sampling

additional contents of ACK packets, each node learns the periodic sampling times of its neighboring nodes, and uses this information to send a shorter wake-up preamble at just the right time. The preamble duration is determined by the potential clock drift since last synchronization. Let T_W be the receiver sampling period, θ the clock drift, and L the interval between communications, then the duration of the preamble T_P need only be:

$$T_P = \min(4L\theta, T_W) \tag{6.1}$$

The packets in WiseMAC also contain a "more" bit (this is also found in the IEEE 802.11 power save protocol), which the transmitter uses to signal to the receiver if it needs to stay awake a little longer in order to receive additional packets intended for it.

6.4.4 Transmitter/receiver-initiated cycle receptions (TICER/RICER)

Similar in spirit to lower-power listening/preamble sampling are the TICER/ RICER techniques [120]. In the transmitter-initiated cycle receiver technique (TICER) (see Figure 6.4(a)), as in low-power listening, the receiver node wakes up periodically to monitor the channel for signals from the sender (which is a wake-up request to send (RTS) signal). The sender sends a sequence of such

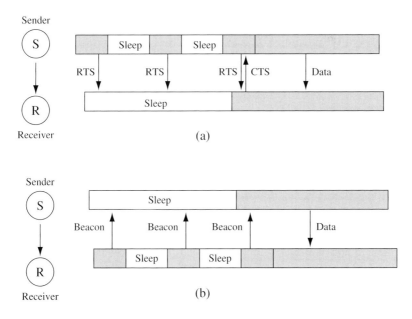

Figure 6.4 Asynchronous sleep using (a) TICER and (b) RICER

RTS signals followed by a short time when it monitors the channel. When the receiver detects an RTS, it responds right away with a CTS signal. If the sender detects a CTS signal in response to its RTS, it begins transmission of the packet. Thus the key difference from preamble sampling is that in TICER the sender sends a sequence of interrupted signals instead of a single long preamble, and waits for an explicit signal from the receiver before transmitting.

In the receiver-initiated cycle receiver technique (RICER), illustrated in Figure 6.4(b), a receiving node periodically wakes up to execute a three phase monitor–send wake-up beacon–monitor sequence. A source that wishes to transmit wakes up and stays in a monitoring state. When it hears a wake-up beacon from a receiver, it begins transmission of the data. The receiver in a monitor state that sees the start of a data packet remains on until the packet reception is completed.

However, one subtlety pertaining to the TICER/RICER techniques is that, while the RTS/CTS/wake-up signals involved are easy to implement at a higher packet level, it can be more challenging to do so at a lower-power RF analog level. This is because to match the transmission to the correct receiver, the receiver needs to uniquely identify itself to the transmitter.

6.4.5 Reconfigurable MAC protocol (B-MAC)

A nice implementation of this class of low-power MAC protocols is the highly reconfigurable and light-weight B-MAC protocol [161]. It provides a limited set of core functionality and an interface that allows the core components to be tuned and configured depending on higher-layer needs. The core of B-MAC consists of the following features which can be turned on/off and used in any combination desired:

- Low-power listening (LPL), which implements the preamble-based wake-up technique described above, to permit nodes to have sleep as the default mode, helping to conserve energy. Different channel sampling durations and preamble durations can be selected by the higher layers.
- Clear channel assessment (CCA), which determines whether the channel is busy or not by examining multiple adjacent samples and using an appropriate outlier detection technique. If CCA is disabled, a scheduling protocol may be implemented above B-MAC. If it is enabled, the backoff duration (in case a busy channel is detected) may be selected by the higher layer. CCA is used for low-power listening.
- Acknowledgements (ACK): If acknowledgements are enabled, a response is sent immediately after receiving any unicast packet.

B-MAC Components				ROM	RAM
Basic B-MAC				3046	166
Basic B-MAC	LPL			4092	170
Basic B-MAC	LPL	ACK		4386	172
Basic B-MAC	LPL	ACK	RTS/CTS	4616	277

Figure 6.5 Components of B-MAC and their memory requirements (in Bytes)

Channel reservation signals for RTS/CTS can be implemented above B-MAC using its control interfaces, but are not part of the core B-MAC itself.

The simplicity, efficiency, and configurability of B-MAC (see Figure 6.5) mean that it is likely to be useful in many practical settings. A desirable feature of B-MAC is that it is designed in a modular fashion and provides core functionality. This enables easy implementation of other more complex MAC protocols *over* B-MAC, including many of the sleep-cycle-based and contention-free time-scheduled protocols that we describe below.

6.5 Sleep-scheduled techniques

6.5.1 Sensor MAC (S-MAC)

The S-MAC protocol [234, 237] is a wireless MAC protocol designed specifically for WSN. As shown in Figure 6.6, it employs a periodic cycle, where each node sleeps a while, and then wakes up to listen for an interval. The duty cycle of this listen–sleep schedule, which is assumed to be the same for all nodes, provides for a guaranteed reduction in energy consumption. During initialization, nodes remain awake and wait a random period to listen for a message providing the sleep–listen schedule of one of their neighbors. If they do not receive such

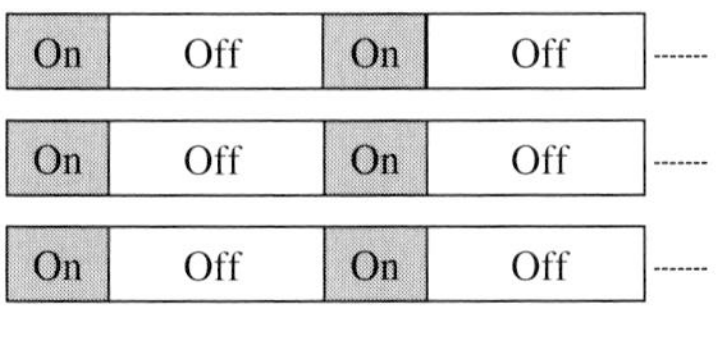

On	Off	On	Off	-------
On	Off	On	Off	-------
On	Off	On	Off	-------

Figure 6.6 Sleep–wake duty cycles in S-MAC

a message, they become synchronizer nodes, picking their own schedules and broadcasting them to their neighbors. Nodes that hear a neighbor's schedule adopt that schedule and are called follower nodes. Some *boundary* nodes may need to either adopt multiple schedules or adopt the schedule of one neighbor (in the latter case, in order to deliver messages successfully, the boundary nodes will need to know all neighbor node schedules.). The nodes periodically transmit these schedules to accommodate any new nodes joining the network. Although nodes must still periodically exchange packets with neighbors for synchronization, this is not a major concern because the listening period is typically expected to be very large (on the order of a second) compared with clock drifts. Sleep schedules are not followed during data transmission. An extension to the basic S-MAC scheme called *adaptive listening* [237] allows the active period to be of variable length, in order to mitigate sleep latency to some extent.

Aside from the sleep scheduling, S-MAC is quite similar to the medium-access contention in IEEE 802.11, in that it utilizes RTS/CTS packets. Both physical carrier sense and the virtual carrier sense based on NAV are employed. S-MAC implements overhearing avoidance, whereby interfering nodes are sent to sleep so long as the NAV is non-zero (the NAV, as in 802.11, is set upon reception of RTS/CTS packets corresponding to the ongoing transmission). S-MAC also provides for fragmentation of larger data packets into several small ones, for all of which only one RTS/CTS exchange is used.

It should be noted that the energy savings in S-MAC come at the expense of potentially significant sleep latency: a packet travelling across the network will need to pause (every few hops, depending on the settings) during the sleep period of intermediate nodes.

6.5.2 Timeout MAC (T-MAC)

The timeout MAC (T-MAC) [37] is a duty-cycled protocol similar in many respects to S-MAC, and like adaptive listening allows modification of the duty cycle. The length of each cycle is kept constant, but the end of the active period is determined dynamically by the use of a timeout mechanism. If a receiver does not receive any messages (data or control) during the timeout interval, it goes to sleep; if it receives such a message, the timer starts afresh after the reception of the message. This renewal mechanism allows for easy adaptation to spatio-temporal variations in traffic.

The basic T-MAC scheme suffers from the so-called *early sleep* problem, which can reduce throughput, particularly in the case of unidirectional flows. When a node has to be silent due to contention in a given cycle, it is unable to send any message to its intended receiver to interrupt its timeout. When the sender can send after the end of the contention period, the intended receiver is already in sleep mode. Two possible solutions to the early sleep problem are proposed and studied in [37]; we mention them only briefly here. The first solution uses an explicit short FRTS (future request to send) control message that can be communicated to the intended recipient asking it to wait for an additional timeout period. The second solution is called "full buffer priority," in which a node prefers sending to receiving when its buffer is almost full. With this scheme, a node has higher priority to send its own packet instead of receiving another packet, and is able to interrupt the timeout of its intended receiver.

6.5.3 Data-gathering MAC (D-MAC)

For packets that need to traverse multiple hops, both S-MAC and T-MAC provide energy savings at the expense of increased delay. This is because the packet can traverse only a few hops in each cycle before it reaches a node that must go to sleep. This is referred to as the *data-forwarding interruption* problem.

An application-specific solution to this problem is provided by the D-MAC (data-gathering MAC) protocol [127], which applies only to flows on a pre-determined data-gathering tree going up from the various network nodes to a common sink. D-MAC essentially applies a staggered sleep schedule, where nodes at each successive level up the tree follow a receive–transmit–sleep sequence that is shifted to the right. These cycles are aligned so that a node at level k is in the receiving mode when the node below it on the tree at level $k + 1$ is transmitting. This is illustrated in Figure 6.7.

The staggered schedule of D-MAC has many advantages – it allows data and control packets (such as requests for adaptive extensions of the active period) to sequentially traverse all the way up a tree with minimum delay; it allows

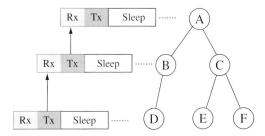

Figure 6.7 The staggered sleep schedule in D-MAC

requests for adaptive extensions of the active period to be propagated all the way up the tree; it reduces interference by separating active periods at the different levels; and it is also shown to reduce the number of nodes that need to be awake when cycle adaption occurs. To deal with contention and interference, D-MAC also includes optional components referred to as *data prediction* and the use of *more-to-send (MTS)* packets.

Despite these advantages, D-MAC in itself is not a general purpose MAC as it applies only to one-way data-gathering trees. However, as we discuss next, the notion of scheduling the wake-up times in such a way as to minimize delay can be extended to other settings.

6.5.4 Delay-efficient sleep scheduling (DESS)

The problem of delay-efficient scheduling of sleep cycles is formulated as a combinatorial optimization problem by Lu *et al.* [128]. The formulation is as follows: each node picks a unique slot out of k slots to use as a reception slot and publishes this to its neighbors. Now, should any of them wish to transmit to this node in any cycle, they wake-up at this slot during the cycle to transmit. On a multi-hop path, the delay that a packet will encounter at each hop is then purely a function of the reception slot times of the corresponding two nodes. For a packet to be sent from node i to its adjacent node j, which have reception slots x_i, x_j respectively, the delay is given as $x_j - x_i \mod k$ if $x_i \neq x_j$, k otherwise. The path delay is the sum of all the per-hop delays along a given path. The end-to-end delay between the two nodes is the minimum path delay. The combinatorial optimization problem of delay-efficient sleep scheduling (DESS) is the following:

Problem 1

Given a communication graph, assign one of k distinct reception slots to each node in the network so that the maximum end-to-end delay between all pairs of nodes in the network (referred to as the *delay diameter* of the network) is minimized.

It has been shown that DESS is an NP-hard problem for arbitrary graphs, but optimal solutions are known in the case of tree- and ring-based graphs and good approximations can be found for a grid network. If nodes are allowed to adopt multiple schedules, then even more significant improvements in delay can be obtained; e.g., on a grid a node can be assigned four schedules of k slots each (one each for transmissions to the left, right, top, and bottom neighbors), which are periodically repeated. On a grid, using multiple schedules can reduce the delay between two nodes that are d hops away on the graph to $O(d + k)$, while the average fraction of wake-up times is still $1/k$. On an arbitrary communication graph with n nodes, the use of two schedules on an embedded tree enables delays that are provably $O((d+k) \cdot \log n)$, while maintaining the same $1/k$ wake–sleep ratio.

6.5.5 Asynchronous sleep schedules

An interesting variation of the problem of scheduling node sleep–wake cycles arises when the need for inter-node synchronization is eliminated [243]. In this case, the design objective is to design independent sleep–wake schedules for individual nodes that guarantee that the wake-up intervals for neighbors overlap. While this requires potentially longer wake-up times, there may be potential gains in ease of design and implementation, particularly in highly dynamic networks, where it is difficult to synchronize schedules across nodes.

Definition 4
A wake-up schedule function (WSF) f_u for a given node u is defined as a binary vector of T slots that indicates the k active slots during which that node u will be awake.

Let the shift-invariant number of overlapping active slots between f_u and f_v be $C(u, v)$. The problem of designing WSFs for all nodes such that they overlap at least m times can be formulated as a combinatorial problem:

Problem 2
For a network graph $G = (V, E)$, given a fixed value of T for the WSF of each node, minimize $\bar{k}$ such that $C(u, v) \geq m$, $\forall (u, v) \in E$. Here $\bar{k} = \frac{\sum_{i \in V} k_i}{|V|}$ is the ensemble average of the number of active slots in each cycle.

This problem can be formulated for asymmetric designs (where different nodes can have different wake-up slot functions) as well as symmetric designs (where all nodes have the same WSF, except for cyclic time shifts). The authors show the following lower bound for the number of active slots in each cycle.

Theorem 7
For any arbitrary WSF design the necessary condition for $C(u, v) \geq m$ is that $k_v \cdot k_u \geq m \cdot T$. For a symmetric WSF design, therefore, the feasible set satisfies $k \geq \sqrt{m \cdot T}$.

Thus, even for a single-slot overlap, the design must have a number of active slots that is at least the square root of the total number of slots in the cycle. The WSF design problem is related to the theory of combinatorial block designs. In particular, a (T, k, m) symmetric block design is equivalent to a WSF symmetric design that has T slots, with k active slots such that m of them overlap between any two of them. It is further shown that techniques from block design theory can be used to compute the desired WSF designs. In particular, it is shown that, if p is a power of a prime, there exists a $(p^2 + p + 1, p + 1, 1)$ design. Figure 6.8 shows an example of a $(7,3,1)$ design.

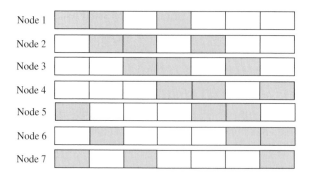

Figure 6.8 A (7,3,1) design for slotted asynchronous wake-up

6.6 Contention-free protocols

Contention-free scheduled medium-access protocols for multi-hop wireless networks date back to the early studies on packet radio networks [162, 169, 29]. These consist of the allocation of time slots for time division multiple-access (TDMA) techniques (or frequency channels for FDMA, distinct codes/hopping sequences for CDMA) to the nodes in the network, while satisfying interference constraints. The commonly used interference constraint (which is admittedly based on a somewhat idealized channel model) requires that no nodes within two hops of each other may use the same slot. The two-hop constraint is needed to prevent hidden terminal collisions. Most of the scheduling schemes that have been proposed and studied in the context of WSN have been TDMA techniques, so we shall largely restrict our attention to these, but FDMA/CDMA-based scheduling techniques involve essentially similar channel assignment problems.

While traditional TDMA assignment techniques do not specifically target energy conservation, it is almost trivial to get energy savings with these techniques because of their predictable communication patterns – all that needs to be done is to ensure that each node is asleep whenever it is not scheduled to transmit or receive.

It is well known that the problem of assigning a minimum number of channels to an arbitrary graph, while ensuring that the two-hop constraints are satisfied, is an NP-hard coloring problem [169]. However, it is possible to provide efficient solutions in practice using heuristic assignments.

One approach to time slot assignment is to perform it in a centralized manner, requiring that a base station gather the full network topology, perform the assignment offline, then distribute it back to the network [31]. However, such solutions do not scale well with network size, particularly in the case of dynamic environments. Decentralized approaches are therefore called for.

6.6.1 Stationary MAC and startup (SMACS)

One decentralized approach is the stationary MAC and startup algorithm proposed in [203]. In this algorithm, each node need only maintain local synchronization. During the starting phase, each node decides on a common communication slot with a neighboring node through handshaking on a common control channel. Each link also utilizes a unique randomly chosen frequency or CDMA frequency-hopping code. It is assumed that there are sufficiently many frequencies/codes to ensure that there are no common frequency/time assignments within interference range, and hence there is no contention. The slot is then used periodically, once each cycle, for communication between the two nodes.

6.6.2 BFS/DFS-based scheduling

One approach to performing channel assignment is the use of breadth-first (BFS) or depth-first (DFS) traversals of a data-gathering tree [3]. In these protocols, every single node is given a unique slot, i.e. there is no spatial reuse of slots. It ensures that each node is given as many slots as the load that it needs to carry. The BFS- and DFS-based slot allocation schemes are illustrated in Figure 6.9.

There are interesting tradeoffs with each approach. With BFS, each node gets contiguous time slots, which is advantageous if the energy costs to switch a radio

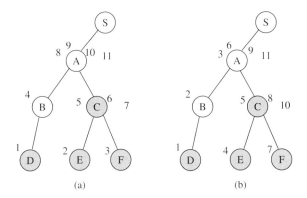

(a)   (b)

Note: Only shaded nodes are sources

Figure 6.9 Time slot allocations for a data-gathering tree: (a) BFS, (b) DFS

from active to sleep and vice versa are large, since it minimizes such transitions at each node. With DFS, each node does not have contiguous slots, but the slots from each sensor source to the sink are contiguous, ensuring that intermediate node buffers are not filled up during data-gathering. This provides low latency to some extent; although the data must remain queued at the sensor node until its transmission slot, once the data is transmitted it is guaranteed to reach the sink in as many steps as the number of hops.

BFS has also been used for more tightly "packed" channel assignment for a data-gathering tree in a scenario where interference constraints are taken into account to provide spatial reuse [205]. In this slot-allocation scheme, each node performs slots assignment sequentially. At each node's turn, it chooses time slots from the earliest possible slot number for its children. Local message exchanges ensure that the slot does not interfere with any already assigned to nodes within two hops of the child. The number of slots to be assigned to each node is pre-determined based on an earlier max–min fair bandwidth allocation phase.

With these techniques, though, it should be kept in mind that they generally require global synchronization.

6.6.3 Reservation-based synchronized MAC (ReSync)

A significant concern with many of the TDMA schemes that provide guaranteed slots to all nodes is that they are not flexible in terms of allowing the traffic from each node to change over time. Reservation-based schemes such as ReSync [32] provide greater flexibility.

In ReSync, each node in the network maintains the notion of an epoch based on its local time alone, but it is assumed that each node's epoch lasts the same duration (or can be synchronized with nearby neighbors accordingly). Each node picks a regular time each epoch based on its local clock to send a short intent message. It is assumed that this selection can be done without significant collisions because the intent message duration is very short. By listening for sufficiently long, each node must further learn when its neighbors send intent messages so that it can wake-up in time to listen to them. When a node wishes to transmit to another node, it indicates in the intent message when it plans to transmit the data (this data transmission time is chosen randomly and indicated as an increment rather than in absolute terms). The intended recipient will then wake-up at the corresponding time (which it can calculate based on its own local clock) in order to receive the message. ReSync does not incorporate an RTS/CTS mechanism to prevent message collisions due to the hidden node problem; however, since the data transmissions are scheduled randomly, any collisions are not persistent.

6.6.4 Traffic-adaptive medium access (TRAMA)

The traffic-adaptive medium access (TRAMA) protocol [167] is a distributed TDMA technique that allows for flexible and dynamic scheduling of time slots. Time epochs are divided into a set of short signaling slots, followed by a set of longer transmission slots. TRAMA consists of three key parts:

1. Neighbor protocol (NP): Nodes exchange one-hop neighboring information during the random access signaling slots. The signaling slots are sufficiently long to ensure that this information is reliably transmitted and all nodes have consistent two-hop neighborhood information.
2. Schedule exchange protocol (SEP): Each node publishes its schedule during its last winning slot in each epoch. The schedule uses bitmaps to efficiently denote the intended recipients (unicast or broadcast) for each of its future packets in the next epoch. If the number of packets a node wishes to send is less than the number of winning slots in an epoch, it gives up the remaining slots (which are packed at the end together between the last used winning slot, referred to as the changeover slot, and the final schedule propagation slot). The changeover slot is used for synchronization; therefore all neighbors are required to wake-up to listen to it. The node sleeps at all times when it is not required to transmit or receive.
3. Adaptive election algorithm (AEA): This algorithm uses a hash function based on node IDs and time to ensure that there is a unique ordering of node priorities among all nodes within any two-hop region at each time. In the simplest version of this algorithm, a node transmits if and only if it has the highest priority among all nodes within its two-hop neighborhood. However, to prevent the waste of a slot, if there is no data to send, TRAMA contains a sophisticated extension that allows for slot reuse.

The slot reuse capability of TRAMA involves the use of additional information from a given node u's point of view. The *absolute winner* for u is the node within two hops of u that has the highest priority; an *alternate winner* for a node u is a node within u's one-hop neighborhood that has the highest priority; the *possible transmitter set* is the set of all one-hop neighbors of u who have highest priority among all their two-hop neighbors that are neighbors of u. A *need transmitter* is the node with the highest priority among u and the possible transmitter set, such that they need additional transmission slots. From a node's perspective, the absolute winner is the assumed transmitter unless the alternate winner is hidden from the absolute winner and is in the possible transmitter set, in which case the assumed transmitter is the alternate winner. Whenever the assumed transmitter gives up, the need transmitter is the true transmitter. Nodes that are not in the schedule listed by the transmitter can shift to sleep mode to save energy, while the relevant transmitter and receiver must stay awake to complete the pertinent communication. It is proved by Rajendran *et al.* [167] that TRAMA is a correct protocol in that it avoids packet losses due to collisions or due to transmission to sleeping nodes.

6.7 Summary

In wireless sensor networks, medium-access protocols provide a dual functionality, not only providing arbitration for access to the channel as in traditional MAC protocols, but also providing energy efficiency by putting the radio to sleep during periods of communication inactivity.

Given the diverse application contexts and the need for tunable tradeoffs in WSN, the B-MAC protocol is a particularly elegant building-block approach to provide basic functionality. B-MAC includes several distinct components including low-power wake-up, clear channel assessment, acknowledgements, and random backoff, that can be individually turned off or tuned by higher layers as desired. Because of its flexibility, sophisticated MAC schemes, including TDMA and sleep-scheduled contention-based schemes, can be easily built on top of it.

Energy efficiency is a key concern for sensor network MAC protocols. As we noted in the previous chapter, significant energy savings are possible by avoiding idle listening. While there have been proposals to use a secondary low-power wake-up radio to achieve this, the simpler low-power listening/preamble sampling technique provides effectively the same benefit. Further savings are possible by setting nodes on periodic sleep–wake cycles, as proposed in the S-MAC technique. S-MAC has been extended to incorporate adaptive listening to reduce the sleep latency caused by periodical wake-up/sleep schedule, while an enhancement to make S-MAC adaptive to traffic variation is addressed in T-MAC. The problem of minimizing end-to-end latency while using sleep modes has been addressed in the D-MAC protocol and the DESS work. Sleep-scheduling techniques that eliminate the need for inter-node synchronization have also been developed.

An important alternative to the above contention-based techniques is TDMA-based MAC protocols. It is trivial in TDMA protocols to avoid idle listening to provide energy efficiency, since all communications in TDMA are pre-scheduled. However, the tradeoff is that TDMA techniques involve higher-complexity distributed algorithms and impose tight synchronization requirements. They are also generally best suited when communication flows are somewhat predictable, although schemes like TRAMA incorporate periodic rescheduling to handle dynamic traffic conditions.

Exercises

6.1 *RTS/CTS:* Are there WSN scenarios in which the RTS/CTS mechanisms are not needed?

6.2 *Preamble sampling:* If a transmitter sends a 20-byte-long preamble on a 10kbps channel, how frequently does a receiver need to sample the medium to ensure that it can wake-up to the receive the packet?

6.3 *S-MAC:* Consider an arbitrarily large square grid of nodes, where each node has exactly four neighbors in the cardinal directions. Consider sleep-scheduling as per the basic S-MAC scheme, where every node is awake in one out of k slots in each cycle, and all nodes in the network are awake at the same slot. Assume that each slot is of sufficient length to transmit exactly one packet (no more) and that traffic is light so that there is no contention. What is the delay incurred in moving a packet from a given node to a destination that is d hops away?

6.4 *DESS:* For a 5×5 grid of sensors propose a delay efficient (DESS) allocation of reception wake-up slots if $k = 5$. How does the delay for this scheme compare with that for the S-MAC style allocation of the same active slot to all nodes?

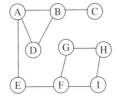

Figure 6.10 A sample graph topology (for exercise 6.6.)

6.5   *Asynchronous slot design:* Give a (13,4,1) design for the slotted asynchronous wake-up scheme.

6.6   *TDMA allocation:* Consider the graph topology in Figure 6.10. Give a minimal assignment of a single time slot to each node, such that no two nodes within two hops of each other share the same slot.

Sleep-based topology control

7.1 Overview

As we discussed in Chapter 2, the core functionality of a wireless sensor network depends on having a network topology that satisfies two important objectives: connectivity and coverage. Connectivity is essential to be able to move movement data between different points within the network, while coverage of the environment ensures that critical phenomena and events can be detected and monitored.

The problem of constructing an initial topology is of course the problem of node deployment. Beyond this, there are many scenarios in which *topology control* is desired. In its broadest sense, topology control can be defined as the process of configuring or reconfiguring a network's topology through tunable parameters after deployment. There are three major tunable parameters for topology control:

1. **Node mobility:** In WSNs consisting of mobile nodes, such as robotic sensor networks, both coverage and connectivity can be adapted by moving the nodes accordingly. We discussed the work that has been done on mobile node deployments and related topology configuration approaches in Chapter 2.
2. **Transmission power control:** In WSNs with static nodes, if the deployment density is already sufficient to guarantee the required level of coverage, the connectivity properties of the network can be adjusted by tuning the transmission power of the constituent nodes. We discussed these kinds of power control-based topology configuration techniques in Chapter 2 as well.
3. **Sleep scheduling:** Finally, consider large-scale static WSNs deployed at a high density, i.e. *over deployed*. In this case, the appropriate topology control

mechanism that provides energy efficiency and extends network lifetime is to turn off nodes that are redundant. The remaining nodes form the active network over which sensing, computation, and communication take place. The challenge is to periodically determine if any active nodes have malfunctioned or if they have suffered energy depletion, so that additional sleeping nodes can be activated if needed. Sleep-based topology control mechanisms are the subject of this chapter.

Certainly, in applications (e.g. in industrial process monitoring) where a limited number of expensive nodes must be deployed in specific locations and must all be active at all times, it is unlikely that redundant nodes will be deployed. Thus sleep-based topology control schemes are not always relevant. However, in other contexts, where (a) the nodes are relatively inexpensive (so that the cost of redundancy is not unreasonably high), (b) the deployed location is remote (so that adding additional nodes at a later time is either undesirable or infeasible), and (c) the precise positioning of each active sensors is not essential, over deploying the network may provide some significant advantages:

1. **Longer lifetime:** The first is that an over-deployed network, if carefully managed through the topology control mechanisms described in this chapter, can have a longer network lifetime. This is obtained by keeping redundant nodes asleep until they are needed to replace the nodes that have failed or depleted their energy.
2. **Robustness:** While one could argue that longer lifetimes could also be obtained for a lower cost merely by adding longer-lifetime batteries, deploying redundant nodes also provides robustness to device failures for reasons other than energy depletion.
3. **Tunable connectivity/coverage:** Another significant advantage of over deploying a network is that it enables the possibility of adaptive topology control techniques that provide tunable levels of connectivity and coverage.

To save energy and maximize lifetime for an over-deployed network, it is essential to ensure that only the minimal set of nodes needed are awake at any time. The goal of sleep-based topology control is to determine which nodes should be awake at a given time to serve both connectivity and coverage needs. Because of the unattended nature of these networks, it is essential that the topology control process be self-configuring and adaptive. In most cases, distributed, localized techniques are essential, to minimize overhead and enable quick reaction to changes.

The basic approach used in many distributed sleep-based topology control techniques for WSN is the following. A set of sleep-related states is defined for each node, and the nodes then transition between these states depending on explicit messages with their neighbors, overheard messages/beacons, or implicit timers.

Consider the following simple example as an illustration. Each node can be in one of three states: sleep, test, and awake. By default a node switches periodically, using a timer, between sleep and test states. During the test state, a local eligibility rule is applied (e.g., "are there less than K awake neighbors?", "is there data intended for me?") to determine whether the node should wake-up. If the rule is satisfied, then the node transitions to the wake state. When the node is an active state, it may (a) remain there forever, if the goal is to ensure that the active backbone of the network remains up at all times, or (b) return back to sleep after some timer expires; for instance, if the eligibility criterion is no longer satisfied.

7.2 Constructing topologies for connectivity

We first describe several sleep-based topology control techniques that have been proposed to address connectivity alone.

7.2.1 Basic/adaptive fidelity energy conserving algorithms (BECA/AFECA)

The basic energy-conserving algorithm (BECA) [229] has three states: sleep, listen, and active. It is illustrated in Figure 7.1(a). The eligibility criterion used to transition from listen to active state is the presence of routing traffic in which the node should participate. Different timers T_s and T_l are used to go from sleep to listen state and back. The node will also transition to sleep from active state if there is no relevant traffic for T_a units. The node can also transition directly from sleep to active state if it has outgoing data to send.

The closely related adaptive fidelity energy-conserving algorithm (AFECA) extends BECA by adapting the sleep time depending on the neighborhood density for increased sleep efficiency. Specifically, the sleep time is chosen as the basic sleep time multiplied by a random number between 1 and N, where N is the number of neighbors estimated in the listening period.

7.2.2 Geographic adaptive fidelity (GAF)

The geographic adaptive fidelity (GAF) technique [230], illustrated in Figure 7.1(b) uses sleep, discovery, and active states. The node moves from the

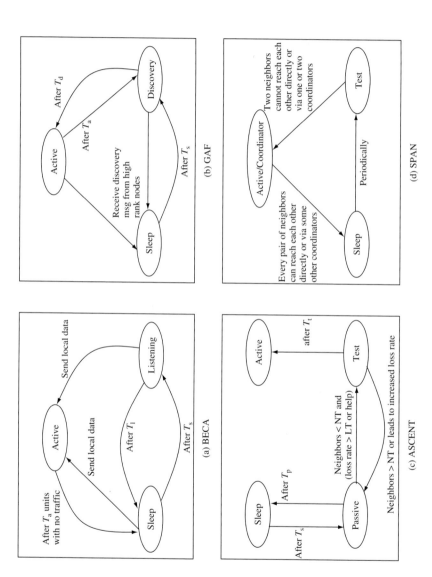

Figure 7.1 State transition diagrams for various sleep-based topology control techniques aimed at connectivity alone

discovery state to the active state and back with timers T_d and T_a respectively; and from the discovery state or the active state to the sleep state if it detects any other higher-priority nodes active within its geographical virtual grid (explained below) through the reception of a discovery message within its neighborhood. It moves from the sleep state to the discovery state after a time T_s. The nodes in discovery/active state transmit discovery messages containing their own ID, the grid ID, and estimated residual lifetime. Nodes with longer residual lifetimes have greater priority.

The goal of GAF is therefore to ensure that a single node with the highest remaining lifetime is awake in each virtual grid. The virtual square grids are created to ensure that all nodes in one grid can communicate with all others in the adjacent grid. This is accomplished by setting the side length of each grid cell r to be sufficiently small so that $r < \sqrt{R/5}$, where R is the radio communication range.

The related cluster-based energy conservation algorithm (CEC) [230] eliminates the need for geographic information necessary to set up the virtual grids. Instead a set of clusters is created such that each has an elected cluster-head (a node is made cluster-head if it has the greatest residual lifetime among its neighbors). To ensure connectivity, additional gateway nodes are elected – primary gateway nodes directly connect two distinct cluster-heads, while secondary gateway nodes connect to other cluster-heads through other secondary/primary gateway nodes. All nodes that are not cluster-heads or gateway nodes are eligible to sleep. The sleep timers are set so that nodes wake-up to run a re-election before the cluster-head's energy is depleted. Nodes that are asleep may wake-up to send their own information, and any data intended for them must be buffered for pick-up after they are awake.

7.2.3 Adaptive self-configuring sensor network topology control (ASCENT)

ASCENT [19] is intended to adapt to highly dynamic environments; nodes wake-up to assist in routing depending on the number of active neighbors and the measured data loss rates in their vicinity.

In ASCENT, illustrated in Figure 7.1(c), nodes are in one of four states: sleep, passive, test, and active. A node in a test state goes to the active state after a timer T_t unless it finds that the number of active neighbors is greater than the threshold NT or that its participation would increase the loss rate further (e.g. due to congestion), in which case it transitions back to the passive state. From the passive state the node transitions to the sleep state after a timer T_p

unless the number of active neighbors is below the desired threshold and either the loss rate is higher than a threshold LT or a help message is received, in which case it transitions to the test state. Help messages are sent by a node that needs the assistance of neighboring nodes to help relay messages to a given destination that it currently is unable to because of high loss. Finally, a node in the sleep state transitions to the passive state after a timer T_s. A key point to note in the basic ASCENT scheme is that nodes do not transition back to sleep from the full active state – this is suitable for applications where nodes in the constructed topology are needed/expected to be active until energy depletion. Message losses are assumed to be detected based on missing packet sequence numbers.

There are a number of tunable parameters in ASCENT that must be adjusted depending on the application needs. The average degree and distance between active nodes needed for connectivity (or even coverage) influence NT and LT. The timers T_t and T_p effect a tradeoff between decision quality and energy efficiency, while T_s provides a tradeoff between energy efficiency and latency of response to failures in the network.

7.2.4 Neighborhood coordinators (SPAN)

SPAN [23] is a related randomized approach to topology control that aims to provide connectivity with redundancy. Only a subset of nodes called coordinator nodes are in active state at any time. Its state transitions are illustrated in Figure 7.1(d). Nodes that are not coordinators go to sleep, waking up periodically to go to the test state to send HELLO messages and check for eligibility to become coordinators. The coordinator eligibility rule is the following: if two neighbors cannot reach each other directly or via one or two coordinators, then the node in test state should become an active coordinator. The decision regarding the eligibility rule is made based on the content of HELLO packets sent by all nodes in which they announce their current coordinators and current neighbors. This algorithm provides for a connected backbone, but not with a minimal set of nodes. It provides sufficient redundancy to ensure that there are coordinators in every radio broadcast range in the network, to minimize congestion bottlenecks.

A randomized prioritized backoff is used to ensure that multiple nodes do not elect to become coordinators simultaneously. Nodes that are likely to benefit a greater set of neighbors are more likely to win contention to become coordinators.

Nodes withdraw as coordinators (i.e. go back from active to sleep state) if they no longer satisfy the eligibility rule, or after some timer period, to ensure load balancing.

7.3 Constructing topologies for coverage

The above-described topology control techniques have focused on waking up the radios of sufficient nodes to maintain good network connectivity. We now turn to the problem of ensuring good coverage in addition to connectivity. Most of these approaches are similar in spirit to those discussed above, except they involve coverage-based eligibility rules for activation of nodes.

7.3.1 Probing environment and adaptive sleep (PEAS)

The probing environment and adaptive sleep (PEAS) technique [235] aims to provide topology control for highly dynamic environments. There are three states in PEAS, as shown in Figure 7.2(a): sleep, probe, and active. From the sleep state the node uses a randomized timer with exponential distribution of rate λ to transition to the probe state. Randomized wake-up times are used to spread the probes from nodes so as to minimize the likelihood that any portion of the network is left without an active node for too long. The rate λ is adapted depending on the environment to ensure that the sleeping rate in every part of the network is about the same desired value λ_d, regardless of spatio-temporal inhomogeneities in node density. In the probing state, a node detects if there is any active node within a probing range R_p, by sending a PROBE message at the appropriate transmit power and listening for REPLY messages. If there are no responses, the node transitions to the active state and stays there until its energy is depleted. If there are REPLY messages, the node computes an updated λ, as described below, and transitions to the sleep state.

   Active nodes measure the rate at which PROBEs are heard from nodes in their neighborhood to estimate the current aggregate neighborhood probing rate $\hat{\lambda}$. The measured probe rate $\hat{\lambda}$ and the desired probe rate λ_d are sent in the REPLY message in response to each PROBE message. The probing node updates its rate λ using the following rule to ensure that the aggregate rate stays around the desired probe rate.

$$\lambda_{new} = \lambda_{current} \frac{\lambda_d}{\hat{\lambda}} \qquad (7.1)$$

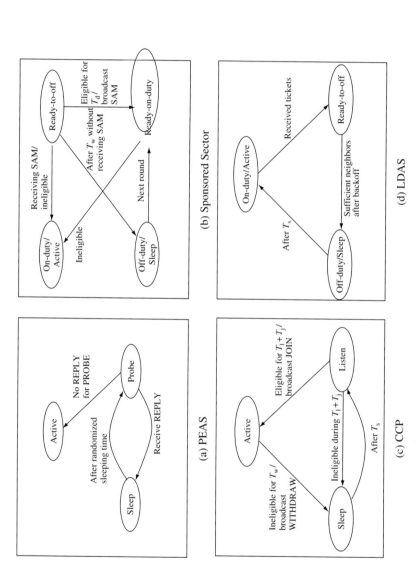

Figure 7.2 State transition diagrams for various sleep-based topology control techniques that address coverage

The choice of λ_d depends on the application tolerance for delays, with the tradeoff being energy efficiency.

Besides connectivity, the PEAS technique also provides tunable coverage. This is done through the eligibility rule in PEAS, where nodes try to ensure that there is an active neighbor within a configurable probing range R_p. R_p could be arbitrarily smaller than the communication range, and allows the density of active nodes to be varied. This does not necessarily guarantee, though, that the original coverage is preserved, which is the goal of the next scheme.

7.3.2 Sponsored sector

Tian and Georganas [210] present an approach that aims to turn off redundant nodes while preserving the original coverage of the network. Each node checks to see if its coverage area is already covered by active neighbors before deciding whether to be on or off. Four states are used: off-duty, ready-on-duty, on-duty, ready-to-off, as seen in Figure 7.2(b).

Nodes that are in the ready-on-duty state investigate whether they are eligible to turn off their sensor and radio, by examining their neighbors' coverage. If they are eligible, they first wait a random backoff time T_d, and broadcast a status advertisement message (SAM), before transitioning to the ready-to-off state. Any neighboring nodes with a longer backoff will not consider these nodes that issued a SAM before them in their neighbor coverage determination. This prevents multiple nodes from shutting off simultaneously. If nodes determine they are ineligible to turn off, they transition to the on-duty state. From the ready-to-off state, after a timer T_w, nodes transition to off-duty unless they receive a SAM from a neighbor and find that they are ineligible to turn off. If they are ineligible to turn off, they transition to on-duty.

The process is conducted via multiple sequential rounds. At the beginning of each round, each node sends a position advertisement message (PAM) to all neighbors within sensing range (by adjusting the transmission power accordingly), containing its sensing range as well as position. The coverage eligibility rule is determined by each node independently in the ready-on-duty state as follows. Based on the PAM messages, each node determines which angular sector of its own sensing range is covered by each of its neighbors (that have not already issued SAMs). If the union of all these *sponsored sectors* is equal to the node's own coverage area, it determines itself to be eligible to turn off. At the end of each round, eligible nodes turn off, while other nodes continue to sense.

7.3.3 Integrated coverage and connectivity protocol (CCP)

Wang *et al.* [220] present an integrated coverage and connectivity protocol (CCP) for sleep-based topology control. This protocol is based on two analytical results we discussed in Chapter 2:

- A convex region is K-covered by a set of sensors if all intersection points between sensing circles of different nodes and between sensing circles and the boundary are at least K-covered.
- A set of nodes that K-cover a given convex region form a K-connected communication graph if $R_c \geq 2R_s$, where R_c, R_s are the communication and sensing ranges respectively.

CCP, shown in Figure 7.2(c), also uses three states: sleep, listen, and active. Sensor nodes transition from the sleep to the listen state after a timer T_s. Nodes in the listen state start a timer with duration T_l. They evaluate their eligibility if they receive a HELLO, WITHDRAW, or JOIN message. A node is eligible if all intersection points of its own circle with those of other sensors or the region boundary are at least K-covered. If the node is eligible, it starts a join timer T_j, otherwise it returns to the sleep state after the timer T_l expires. If a node hears a JOIN beacon from a neighbor after the join timer is started it becomes ineligible and cancels the join timer T_j and goes to sleep. If the join timer is not cancelled, when it expires the node broadcasts a JOIN beacon and enters the active state. In active state, a node periodically sends HELLO messages. An active node that receives a HELLO message checks its own eligibility; if ineligible, it starts a withdraw timer T_w after which it broadcasts a WITHDRAW message and goes to sleep. If it becomes eligible before the withdraw timer expires (due to the reception of a WITHDRAW or HELLO message), it cancels the withdraw timer and remains in active state.

If $R_c \geq 2R_s$, since K-coverage guarantees K-connectivity, the protocol inherently provides both coverage and connectivity guarantees. However, for the case where $R_c < 2R_s$, the authors create a hybrid technique merging CCP with SPAN. This is done by simply combining the eligibility rules for CCP as well as SPAN together, so that both coverage and connectivity concerns are integrated in determining which nodes should form the active topology.

7.3.4 Lightweight deployment-aware scheduling (LDAS)

The LDAS protocol [228] uses a random-voting technique for topology control. The key objective of this technique is to provide probabilistic coverage guarantees without requiring exact location information from neighboring nodes.

There are three states in LDAS as shown in Figure 7.2(d): on-duty (active), ready-to-off, and off-duty. During the on-duty state, the sensor node checks the number of its working neighbors n. If this is larger than the required threshold r (described below), then it sends out penalty tickets to $n - r$ randomly selected active neighbors. A node that receives greater than a threshold number of tickets b goes to the ready-to-off state. In this state a random backoff time between 0 and W_{max} is used as a timer. If it still has sufficient neighbors at the end of this state, it goes to the off-duty state, erases all tickets, and stays asleep for a timer of duration T_s.

It is assumed that all nodes are placed uniformly in an area with average node density of n. Each node sends a ticket to each neighbor with probability $\frac{n-r}{n}$ if $r \leq n$. The ticket threshold b needs to be chosen so that the remaining average number of nodes after the removal of all nodes with b or more tickets is r. It is hard to estimate b in general, but in low-density settings, it is shown that

$$b \approx (n - r) - \sqrt{2(n - r) \ln \frac{n + 1}{r}} \tag{7.2}$$

The goal of this algorithm is to ensure that r neighbors are kept awake. Under the assumption of uniform random deployment, bounds are derived in [228] that relate the probability of complete redundancy as well as the average partial redundancy to the number of neighbors. In particular, it is shown that only five neighbors are necessary for each sensor's 90% sensing area to be covered by its neighbors, while having 11 neighbors guarantees, with greater than 90% probability, that its range can be completely covered by its neighbors. These constant numbers of neighbors for coverage that are independent of the total number of nodes in the network N, of course, have to be balanced with the requirement that $O(\log N)$ neighbors are required to ensure network-wide connectivity with high probability, as per the results of Xue and Kumar [232].

7.4 Set K-cover algorithms

A completely different approach to the problem of sleep-based topology control for sensor coverage is to formulate it as the following SET K-COVER problem [202]:

Problem 3
Given a collection $C = \{S_j\}$ of subsets of a set S, and a positive integer K. Does C contain K disjoint covers of the set S, i.e. covers $C_1, C_2, \ldots, C_K$, where $C_i \subseteq C$ such that every element of S belongs to at least one member of each of C_i?

This problem is of relevance to sleep-based topology control, as it essentially is the problem of selecting K disjoint sets of nodes from the network, such that each set provides full coverage of the region. Once these are found, the first covering set of sensors is then activated with all other nodes kept in sleep state, then these are put to sleep, and the second covering set is activated, and so on in a successive series of rounds. The larger K is, the longer the lifetime of the network. The problem is NP-complete, but a centralized heuristic technique to maximize the number of covers is provided in [202].

A slightly different version of this problem has also been studied, where the goal is to identify a partition ζ of the subsets into K-covers that maximize the sum of the total coverage provided by each cover (i.e. maximize $N(\zeta) = \sum_i^k |\bigcup_{S_j \in C_i} S_j|$). This formulation is more relaxed yet more reasonable in that it is not required that each cover provide full coverage of the area. Three algorithms have been considered for this problem [2]: (i) a naive greedy algorithm in which each node randomly assigns itself to one of the covers, (ii) a distributed algorithm in which each sensor adds itself in turn to the cover to which it adds the most additional coverage, and (iii) a centralized greedy algorithm, which is the same as the distributed approach except the additional coverage area is further weighed by likelihood of coverage by other nodes later in the process. Analytical performance bounds are obtained; a surprising result is that in expectation, the simple randomized algorithm can provide an approximation within $1 - \frac{1}{e}$ of the optimal solution, while the distributed and centralized algorithms provide deterministic approximations of 0.5 and $1 - \frac{1}{e}$ respectively. It is shown that for highly dense networks the energy savings are proportional to the density of the network, and excellent performance is possible in terms of both coverage and lifetime.

7.5 Cross-layer issues

It is crucial to clarify the relationship between sleep-based topology control with respect to medium-access and routing techniques.

1. **Relationship to routing:** Recall that the goal of topology control mechanism is to provide the network substrate over which routing can take place. Hence it is clear immediately that topology control must be implemented below the routing layer. Specifically, the routing protocol must only consider nodes that are in the active state as per the topology control.

2. **Relationship to MAC:** The relationship between topology control and medium-access protocols is somewhat more complicated. Recall from Chapter 6 that WSN MAC protocols perform two functions: they provide not only the traditional function of arbitrating access to the radio medium, but also energy savings by keeping nodes asleep, either whenever they are not communicating or on some form of a periodic schedule.

   Clearly, since topology control mechanisms themselves need to be aware of neighbors and to exchange messages with them, they must be implemented above the MAC layer. However, a source of confusion is that both the MAC layer and topology control mechanisms put node radios to sleep. We limit our discussion here to radio sleep alone, since MAC protocols do not necessarily put sensors or processors to sleep.

   The transitions of nodes between states in the topology control mechanism take place over much longer time scales than the MAC sleep/wake-up modes. MAC-controlled sleep periods are therefore best invoked only once the node has already been activated to be part of the topology.

   Consider the following five modes (states), which can exist when both MAC and topology control co-exist:

   1. MAC-induced sleep mode (MS)
   2. MAC-induced active mode (MA)
   3. Topology control-induced sleep mode (TCS)
   4. Topology control-induced test mode (TCT)
   5. Topology control-induced active mode (TCA)

   These modes need to be coordinated to ensure that they do not conflict with each other. When the topology control protocol puts a node to sleep (TCS mode), the MAC-layer sleep/wake transitions must be completely disabled. Hence any MAC-layer periodic sleep or wake-up should not be performed in this mode. In most cases, the topology control-induced test mode (TCT) should also be treated similarly, i.e. with the MAC sleep–wake control disabled. When the topology control puts a node into the active mode (TCA), the MAC-induced sleep and active modes (MS, MA) are both possible. When the node is in TCA and MA, then the radio is on and working. When the node is in TCA and MS, then even though the node is an active participant in the network topology, the radio is in a MAC-induced sleep mode. On the whole, therefore, it must be ensured that the MAC has full control over radio sleep/wake-up in the TCA mode, and no control at all in the TCS/TCT modes.

7.6 Summary

In WSN applications involving remote surveillance with inexpensive nodes it may be reasonable to over-deploy the network with higher density than needed for basic operation. To extend the lifetime of operation in an over-deployed network, it is necessary to have some mechanism to keep redundant nodes inactive until the active nodes in their neighborhood fail or deplete their energy. This is the key functionality of the sleep-based topology control mechanisms examined in this chapter.

The topology control mechanism must ensure that the network provides adequate connectivity and coverage, despite node failures, so long as sufficient redundancy is available. For unattended operations, it is essential that the topology control mechanism be distributed in nature. Most of the techniques we described in this chapter essentially operate in a similar manner: each node uses local rules and observations to transition between sleep, test, and active states (see Figure 7.1 and Figure 7.2). Some of these techniques focus exclusively on connectivity, such as BECA/AFECA, GAF, CEC, ASCENT, and SPAN, while others, such as PEAS, the sponsored-sector technique, CCP, and LDAS, address coverage issues as well (either in isolation or integrated with connectivity). The set K-cover approaches offer another alternative.

A subtle design issue is that sleep-based topology control protocols must co-exist with any sleep-based MAC protocols that may be implemented on the same network. MAC-level sleep–wake control should be activated only when the node is an active participant of the topology.

Exercises

7.1 *State timeline:* Consider a node in a network running the BECA topology control mechanism. The three timer parameters are set as follows: $T_s = 10\,\mathrm{s}$, $T_1 = 2\,\mathrm{s}$, $T_a = 3\,\mathrm{s}$. Assume the node just starts on its sleep mode at time $0\,\mathrm{s}$. There is routing traffic in the network that it can potentially participate in at times $5\,\mathrm{s}$, $11\,\mathrm{s}$, $13\,\mathrm{s}$, and the node has some local data of its own to send at time $19\,\mathrm{s}$. Given this information, draw a time line showing periods when the node is asleep, when it is in listen mode, and when it is in active mode.

7.2 *GAF virtual grid setting:* Prove why setting the side of the virtual grid in GAF to be $r < \sqrt{R/5}$ suffices to ensure that all nodes in adjacent grids (in the four directions) can communicate with each other. What is the bound

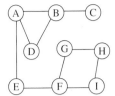

Figure 7.3 A sample communication network graph (for exercise 7.3.)

on r if it is required that nodes should also be able to communicate with neighbors in diagonally adjacent grids (i.e. with the nearest eight grids)?

7.3 *SPAN:* For the communication network graph shown in Figure 7.3, identify a set of nodes such that the nodes could be coordinators when using the SPAN topology control mechanism.

7.4 *PEAS:* Consider a 4×4 grid deployment of 16 sensor nodes with coordinates (0, 0), (0, 1), ..., (3, 3). Let us assume that 25% of the deployed nodes are required to wake-up within the first minute for the network to function properly.

(a) Based on the fact that each node sleeps for an exponential distributed duration with probability density function $f(t_s) = \lambda e^{-\lambda t_s}$, what is the desired value of λ_d to ensure that 4 nodes wake-up at least once within the first minute?

(b) Assume that the initial λ of all nodes is set to 0.012 and the probing range $R_p = 1$, what is the expected new value of λ after one round of adaptation based on equation (7.1)? If $R_p = 2$, what is the expected new value of λ?

(c) Simulate the protocol by initializing λ of all nodes as a random variable between 0 to 0.1. Draw the curve of the aggregated λ within 5 minutes with $R_p = 1$ and $R_p = 2$, respectively.

7.5 *Sponsored sector:* Simulate the deployment of nodes one by one randomly in a unit area, with sensing range circles of $R_s = 0.2$, determining whether to activate each one based on the sponsored-sector technique. Plot the number of active nodes with respect to the total number of nodes. At what point does this curve become saturated?

7.6 *LDAS:* Essentially, the purpose of the LDAS topology control mechanism is to keep the average number of active nodes in a neighborhood at r. Consider a simple heuristic where any sensor node goes to sleep with probability $\frac{n-r}{r}$, where n is the number of its working neighbors. Simulate this heuristic in a network with initial average number of neighbors varing from 10 to 30, and compare with a simulation of LDAS based on equation (7.2).

7.7 *Set K-cover:* Consider the simplest randomized Set *K*-cover algorithm, in which each node allocates itself at random to one of the *K*-covers. For evaluation, the following coverage metric is to be used: the maximum distance between any active node in the network and its nearest active neighbor, averaged over each cover (note that it is desirable to minimize this metric for superior coverage). Perform simulations in which n nodes are deployed randomly with uniform distribution in a unit area, varying K, showing the performance of the simple randomized algorithm in terms of this maximum distance metric.

8

Energy-efficient and robust routing

8.1 Overview

Information routing in wireless sensor networks can be made robust and energy-efficient by taking into account a number of pieces of state information available locally within the network.

1. **Link quality:** As we discussed in Chapter 5, link quality metrics (e.g. packet reception rates) obtained through periodic monitoring are very useful in making routing decisions.
2. **Link distance:** Particularly in case of highly dynamic rapidly fading environments, if link monitoring incurs too high an overhead, link distances can be useful indicators of link quality and energy consumption.
3. **Residual energy:** In order to extend network lifetimes it may be desirable to avoid routing through nodes with low residual energy.
4. **Location information:** If relative or absolute location information is available, geographic routing techniques may be used to minimize routing overhead.
5. **Mobility information:** Recorded information about the nearest static sensor node near a mobile node is also useful for routing.

We examine in this chapter several routing techniques that utilize such information to provide energy efficiency and robustness.

8.2 Metric-based approaches

Robustness can be provided by selecting routes that minimize end-to-end retransmissions or failure probabilities. This requires the selection of a suitable metric for each link.

8.2.1 The ETX metric

If all wireless links are considered to be ideal error-free links, then routing data through the network along shortest hop–count paths may be appropriate. However, the use of shortest hop–count paths would require the distances of the component links to be high. In a practical wireless system, these links are highly likely to be error-prone and lie in the transitional region. Therefore, the shortest hop–count path strategy will perform quite poorly in realistic settings. This has been verified in a study [39], which presents a metric suitable for robust routing in a wireless network. This metric, called ETX, minimizes the expected number of total transmissions on a path. Independently, an almost identical metric called the minimum transmission metric was also developed for WSN [225].

It is assumed that all transmissions are performed with ARQ in the form of simple ACK signals for each successfully delivered packet. Let d_f be the packet reception rate (probability of successful delivery) on a link in the forward direction, and d_r the probability that the corresponding ACK is received in the reverse direction. Then, assuming each packet transmission can be treated as a Bernoulli trial, the expected number of transmissions required for successful delivery of a packet on the link is:

$$ETX = \frac{1}{d_f \cdot d_r} \tag{8.1}$$

This metric for a single link can then be incorporated into any relevant routing protocol, so that end-to-end paths are constructed to minimize the sum of ETX on each link on the path, i.e. the total expected number of transmissions on the route.

Figure 8.1 shows three routes between a given source A and destination B, each with different numbers of hops with the labelled link qualities (only the forward probabilities are shown, assume the reverse probabilities are all $d_r = 1$).

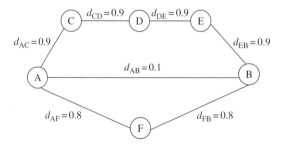

Figure 8.1 Illustration of the ETX routing metric

The direct transmission from A to B incurs 10 retransmissions on average. The long path through nodes C, D, and E incurs 1.11 retransmissions on each link, hence a total of 4.44 retransmissions. The third path, through node F, incurs 1.25 retransmissions on each link for a total of 2.5 retransmissions. This is the ETX-minimizing path. This example shows that ETX favors neither long paths involving a large number of short-distance (high-quality) links, nor very short paths involving a few long-distance (low-quality) links, but paths that are somewhere in between.

The ETX metric has many advantages. By minimizing the number of transmissions required, it improves bandwidth efficiency as well as energy efficiency. It also explicitly addresses link asymmetry by measuring reception probabilities in both directions.

One of the practical challenges is in determining the values of the packet reception probabilities d_f and d_r, by performing an appropriate link monitoring procedure. This can be done through some form of periodic measurement using sliding windows.

8.2.2 Metrics for energy–reliability tradeoffs (MOR/MER)

If the environment contains highly mobile objects, or if the nodes are themselves mobile, the quality of links may fluctuate quite rapidly. In this case, use of ETX-like metrics based on the periodic collection of packet reception rates may not be useful/feasible.

Reliable routing metrics for wireless networks with rapid link quality fluctuations have been derived analytically [106]. They explicitly model the wireless channel as having multi-path fading with Rayleigh statistics (fluctuating over time), and take an outage probability approach to reliability. Let d represent the distance between transmitter and receiver, η the path-loss exponent, SNR the normalized signal-to-noise ratio without fading, f the fading state of the channel, then the instantaneous capacity of the channel is described as:

$$C = \log\left(1 + \frac{|f|^2}{d^\eta} SNR\right) \tag{8.2}$$

The outage probability P_{out} is defined as the probability that the instantaneous capacity of the channel falls below the transmission rate R. It is shown that

$$P_{out} = 1 - \exp\left(\frac{-d^\eta}{\mu SNR^*}\right) \tag{8.3}$$

where $SNR^* = SNR/(2^R - 1)$ is a normalized SNR, and $\mu = E[|f|^2]$ is the mean of the Rayleigh fading. Based on this formulation, the authors derive the following results for the case that each transmission is limited to the same power:

Theorem 8

Let the end-to-end reliability of a given route be defined as the probability that none of the intermediate links suffer outage. Then, assuming that each link has the same transmitted signal-to-noise ratio, the most reliable route between two nodes is one that minimizes the path metric $\sum_i d_i^\eta$, where d_i is the distance of the ith hop in the path.

This metric (d^η for each link of distance d) is referred to as the minimum outage route (MOR) metric.

They also derive the following result for the case with power control:

Theorem 9

The minimum energy route between nodes that guarantees a minimum end-to-end reliability $R_{\min}$ is the route which minimizes the path metric $\sum_i \sqrt{d_i^\eta}$.

In this case the metric for each link is $\sqrt{d^\eta}$ with a proportional power setting, which is referred to as the minimum energy route (MER) metric. It should be noted, however, that here the energy that is being minimized is only the distance-dependent output power term – not the cost of receptions or other distance-independent electronics terms. It turns out that the MER metric can also be used to determine the route which maximizes the end-to-end reliability metric, subject to a total power constraint.

One key difference between MOR/MER metrics and the ETX metric is that they do not require the collection of link quality metrics (which can change quite rapidly in dynamic environments), but assume that the fading can be modelled by a Rayleigh distribution. Also, unlike ETX, this work does not take into account the use of acknowledgements.

8.3 Routing with diversity

Another set of techniques exploits the diversity provided by the fact that wireless transmissions are broadcast to multiple nodes.

8.3.1 Relay diversity

The authors of [106] also propose and analyze the reliability–energy tradeoffs for a simple technique for providing diversity in wireless routing that exploits the wireless broadcast advantage. This is illustrated in Figure 8.2 by a simple two-hop route from A to B to C. With traditional routing, the reliability of the

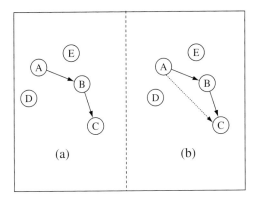

Figure 8.2 Illustration of relay diversity: (a) traditionally C receives a packet from A successfully only if the transmission from A to B and B to C are both successful; (b) with relay diversity, the transmission could also be successful in addition if the transmission from A to B is overheard successfully by C

path is purely a function of whether transmissions on A and B and B and C were both successful. However, if C is also allowed to accept packets directly from A (whenever they are received without error), then the reliability can be further increased without any additional energy expenditure.

Allowing such packet receptions within two hops, it is shown that in the high-SNR regime, the end-to-end outage probability decays as $(SNR)^{-2}$ (a second-order diversity gain). The authors further conjecture that, when nodes within L hops can communicate with each other with high SNR, the end-to-end outage probability would decay as SNR^{-L}.

One tradeoff in using this technique is that it requires a larger number of receivers to be actively overhearing each message, which may incur a radio energy penalty.

8.3.2 Extremely opportunistic routing (ExOR)

A related innovative network layer approach to robust routing that takes unique advantage of the broadcast wireless channel for diversity is the extremely opportunistic routing (ExOR) technique [10]. Unlike traditional routing techniques, in ExOR the identity of the node, which is to forward a packet, is not pre-determined before the packet is transmitted. Instead, it ensures that the node closest to the destination that receives a given packet will forward the packet further. While this technique does not explicitly use metric-based routing, the protocol is designed to minimize the number of transmissions as well as the end-to-end routing delay (Figure 8.3).

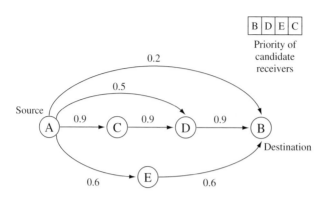

Figure 8.3 ExOR routing

In detail, the protocol has three component steps:

1. Priority ordering: At each step, the transmitter includes in the packet a schedule describing the candidate set of receivers and the priority order in which
   they should forward the packet. The candidate list and ordering are constructed to ensure high likelihood of forward progress, as well as to ensure
   that receivers that lie on shorter routes to the destination (which are also
   generally the closest to the destination) have higher priority.
2. Transmission acknowledgements: A distributed slotted MAC scheme is used
   whereby each candidate receiver sends the ID of the highest-priority successful recipient known to it. All nodes listen to all ACK slots. This constitutes a
   distributed election procedure whereby all nodes can determine which node
   has the highest priority among the candidates that received the packet successfully. Even if a node does not directly hear the highest-priority node's
   ACK, it is likely to hear that node's ID during another candidate's ACK.
   Thus, by transmitting the IDs as per this protocol, and having all nodes wait
   to receive all ACKs, an attempt is made to suppress duplicate forwarding to
   the greatest extent possible.
3. Forwarding decision: After listening to all ACKs, the nodes that have not
   heard any IDs with priorities greater than their own will transmit. There is a
   small possibility that ACK losses can cause duplicate transmissions; these are
   further minimized to some extent by ensuring that no node will ever retransmit
   the same packet twice (upon receiving it from different transmitters).

There are several nice features of the ExOR protocol. The chief among these is
that it implements a unique network-layer diversity technique that takes advantage
of the broadcast nature of the wireless channel. Nodes that are further away

(yet closer to the destination) are less likely to receive a packet, but, whenever they do, they are in a position to act as forwarders. The almost counter-intuitive approach of routing without pre-specifying the forwarding node thus saves on expected delay as well as the number of transmissions.

As with the relay diversity technique described before, ExOR also requires a larger set of receivers to be active, which may have an energy penalty. Moreover, to determine the priority ordering of candidate receivers, the inter-node delivery ratios need to be tracked and maintained.

8.4 Multi-path routing

One basic solution for robustness that has been proposed with several variations is the use of multiple disjoint or partially disjoint routes to convey information from source to destination. There is considerable prior literature on multi-path routing techniques; we will highlight here only a few recent studies focused on sensor networks.

8.4.1 Braided multi-path routing

A good example of the use of localized enforcement-based routing schemes, such as Directed Diffusion to provide for multi-path robustness, is the "braided multi-path" schemes for alternate path routing in wireless sensor networks [58]. In alternate path routing, there is a primary path that is mainly used for routing and several alternate paths are maintained for use when a failure occurs on the primary path. A braided path is one where node disjointedness between the alternate paths is not a strict requirement. It is defined as one in which for each node on the main path there exists an alternate path from the source to the sink that does not contain that node, but which may otherwise overlap with the other nodes on the main path.

The braided multi-path construction is compared with a localized disjoint multi-path construction in the study [58] in terms of resiliency to both isolated and pattern failures, as well as the overhead required for multi-path maintenance (which is assumed to be proportional to the number of participating nodes). While the disjoint multi-path techniques have greater resiliency to pattern failures than the braided multi-path techniques (which suffer due to the geographic proximity of all alternate paths), this comes at the expense of significantly greater overhead. Particularly for isolated failures, the braided approach can be significantly more resilient and energy-efficient.

8.4.2 Gradient cost routing (GRAd)

The gradient routing technique (GRAd) [163], provides a simple mechanism for multi-path robustness. All nodes in the network maintain an estimated cost to each active destination (this set would be restricted to the sinks in a sensor network). In the simplest case, the cost metric would be just the number of hops; however, the protocol can be enhanced to handle other metrics. When a packet is transmitted, it includes a field that indicates the cost it has accrued to date (i.e. number of hops traversed), and a remaining value field, that acts as a TTL (time-to-live) field for the packet. Any receiver that receives this packet and notes that its own cost is smaller than the remaining value of the packet can forward the message, so long as it is not a duplicate. Before forwarding, the accrued cost field is incremented by one and the remaining value field is decremented by one (in the case of hop-count metric, the increment/decrements would be accordingly different for other metrics). One issue that needs to be taken into account in practice is that in the basic GRAd scheme cost fields are established using a reverse-path approach, which assumes the existence of bidirectional links. Since GRAd allows multiple nodes to forward the same message, it acts essentially as a limited directed flood and provides significant robustness, at the cost of larger overhead.

8.4.3 Gradient Broadcast routing (GRAB)

The Gradient Broadcast mechanism (GRAB) [236] enhances the GRAd approach by incorporating a tunable energy–robustness tradeoff through the use of credits. Similar to GRAd, GRAB also maintains a cost field through all nodes in the network. The packets travel from a source to the sink, with a credit value that is decremented at each step depending on the hop cost. An implicit credit-sharing mechanism ensures that earlier hops receive a larger share of the total credit in a packet, while the later hops receive a smaller share of the credit. An intermediate forwarding node with greater credit can consume a larger budget and send the packet to a larger set of forwarding eligible neighbors. This allows for greater spreading out of paths initially, while ensuring that the diverse paths converge to the sink location efficiently. This is illustrated in Figure 8.4, which shows the set of nodes that may be used for forwarding between a given source and the sink.

The GRAB-forwarding algorithm works as follows. Each packet contains three fields: (i) R_0 – the credit assigned at the originating node; (ii) C_0 – the cost-to-sink at the originating node; and (iii) U – the budget already consumed from the source to the current hop. The first two fields never change in the packet, while

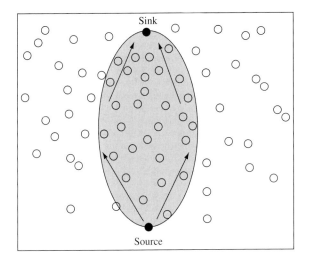

Figure 8.4 Forwarding mesh for gradient broadcast routing

the last is incremented at each step, depending on the cost of packet transmission (e.g., it could be related to the power setting of the transmission). To prevent routing loops, only receivers with lower costs can be candidates for forwarding. Each candidate receiver i with a cost-to-sink of C_i computes a metric called β and a threshold θ as follows:

$$\beta = 1 - \frac{R_{oi}}{R_o} \tag{8.4}$$

$$\theta = \left(\frac{C_i}{C_o}\right)^2 \tag{8.5}$$

where

$$R_{oi} = U - (C_o - C_i) \tag{8.6}$$

The expression R_{oi} determines how much credit has already been used up in traversing from the origin to the current node. The metric β, therefore, is an estimate of the remaining credit of the packet. The threshold θ is a measure of remaining distance to the sink. The candidate node will forward the message so long as $\beta > \theta$. The square gives the θ threshold a weighting, so that the threshold is more likely to be exceeded in the early hops than the later hops, as desired. The authors of GRAB show that the choice of initial credit, R_o, provides a tunable parameter to increase robustness at the expense of greater energy consumption.

8.5 Lifetime-maximizing energy-aware routing techniques

A number of studies have explored the issue of energy aware, lifetime-maximizing routing approaches for wireless ad hoc and sensor networks. Many of these are based on identifying and defining suitable shortest-path link metrics, while some derive energy-efficient routes for a network using a global optimization formulation.

8.5.1 Power-aware routing

In an ideal, lightly loaded environment, assuming all links require the same energy for the transmission of a packet, the traditional minimum hop-count routing approach will generally result in minimum energy expended per packet. If different links have uneven transmission costs, then the route that minimizes the energy expended in end-to-end delivery of a packet would be the shortest path route computed using the metric $T_{i,j}$, the transmission energy for each link i, j. However, in networks with heterogeneous energy levels, this may not be the best strategy to extend the network lifetime (defined, for instance, as the time till the first node exhaustion).

The basic power-aware routing scheme [200] selects routes in such a way as to prefer nodes with longer remaining battery lifetime as intermediate nodes. Specifically, let R_i be the remaining energy of an intermediate node i, then the link metric used is $c_{i,j} = \frac{1}{R_i}$. Thus, the path P (indicating the sequence of transmitting nodes for each hop) selected by a shortest-cost route determination algorithm (such as Dijkstra's or Bellman-Ford) would be one that minimizes $\sum_{i \in P} \frac{1}{R_i}$.

8.5.2 Lifetime-maximizing routing

While minimizing per-hop transmission costs minimizes total energy, avoiding nodes with low residual energy prevents early node failure. However, considering these goals separately, as in the above, may not optimize the system lifetime. What is needed is a technique that balances the two goals, selecting the minimum energy path when all nodes have high energy at the beginning, and avoiding the low residual energy nodes towards the end.

Chang and Tassiulas [20] propose the following link metric, which is a function of the transmission cost on the link $T_{i,j}$, the residual energy of the transmitting node R_i, and the initial energy of the transmitting node E_i:

$$c_{i,j} = T_{i,j}^a R_i^{-b} E_i^c \qquad (8.7)$$

This general formulation captures a wide range of metrics. If $(a, b, c) = (0, 0, 0)$, we have a minimum hop metric; if $(a, b, c) = (1, 0, 0)$, we have the minimum energy-per-packet metric; if $b = c$, then normalized residual energies are used, while $c = 0$ implies that absolute residual energies are used; if $(a, b, c) = (0, 1, 0)$, we have the inverse-residual-energy metric suggested in [200]. However, simulation results in [20] suggest that a non-zero a and relatively large $b = c$ terms provide the best performance (e.g. (1, 50, 50)).

8.5.3 Load-balanced energy-aware routing

In order to provide an additional level of load balancing, which is necessary in a static network, the following cost-based probabilistic forwarding mechanism [189] can be used. Nodes forward packets only to neighbors that are closer to the destination. Let the cost to destination from node i through a candidate neighbor j be given as $C_{ij} = C_j + c_{ij}$, where C_j is the expected minimum cost-to-destination from j and c_{ij} is any link cost metric (e.g. the $T_{i,j}^a R_i^{-b} E_i^c$ metric discussed above). For each neighbor j in its set of candidate neighbors (N_i), node i assigns a forwarding probability that is inversely proportional to the cost to destination

$$P_{i,j} = \frac{C_{ij}^{-1}}{\sum_{k \in N_i} C_{ik}^{-1}} \tag{8.8}$$

Node i then calculates the expected minimum cost to destination for itself as

$$C_i = \sum_{j \in N_i} P_{ij} C_{ij} \tag{8.9}$$

Each time the node needs to route any packet, it forwards to any of its neighbors randomly with the corresponding probability. This provides for load balancing, preventing a single path from rapidly draining energy.

8.5.4 Flow optimization formulations

Chang and Tassiulas [21] also formulate the global problem of maximizing the network lifetime with known origin–destination flow requirements as a linear program (LP), and propose a flow augmentation heuristic technique based on iterated saturation of shortest-cost paths to solve it. The basic idea is that in each iteration every origin node computes the shortest cost path to its destination, and augments the flow on this path by a small step. After each step the costs are recalculated, and the process repeated until any node runs out of its initial energy E_i.

We should note that such LP formulations have been widely used by several authors in the literature to study performance bounds and derive optimal routes. Bhardwaj *et al.* use LP formulations to derive bounds on the lifetime of sensor networks [9]. LP-based flow augmentation techniques are used by Sadagopan and Krishnamachari [179] for a related problem involving the maximization of total data gathered for a finite energy budget. Techniques to convert multi-session flows obtained from such linear programming formulations into single-session flows are discussed in [151]. Kalpakis *et al.* [99] also present an integer flow formulation for maximum lifetime data-gathering with aggregation, along with near-optimal heuristics. Ordonez and Krishnamachari present non-linear flow optimization problems that arise when variable power-based rate control is considered, and compare the gains obtained in energy-efficient data-gathering with and without power control [110].

To illustrate this approach, consider the following simple flow-based linear program. Let there be n-numbered source nodes in the network, and a sink labelled $n+1$. Let f_{ij} be the data rate on the corresponding link, C_{ij} the cost of transmitting a bit on the link, R the reception cost per bit at any node, T the total time of operation under consideration, E_i the available energy at each node, and B_i the total bandwidth available at each node.

$$\max \quad \sum_{j=i}^{n} f_{i,n+1} \cdot T$$

$$\text{s.t. } \forall i \neq (n+1), \ \sum_{j=1}^{n+1} f_{ij} - \sum_{j=1}^{n} f_{ji} \geq 0 \qquad \text{(a)}$$

$$\left(\sum_{j=1}^{n+1} f_{ij} \cdot C_{ij} + \sum_{j=1}^{n} f_{ji} \cdot R \right) \cdot T \leq E_i \qquad \text{(b)}$$

$$\sum_{j=1}^{n+1} f_{ij} + \sum_{j=1}^{c} f_{ji} \leq B_i \qquad \text{(c)}$$

This linear program maximizes the total data gathered during the time duration T. It incorporates (a) a flow conservation constraint, (b) a per-node energy constraint, and (c) a shared bandwidth constraint.

8.6 Geographic routing

Since location information is often available to all nodes in a sensor network (if not directly, then through a network localization algorithm) in order to provide location-stamped data or satisfy location-based queries, geographic routing techniques [135] are often a natural choice.

8.6.1 Local position-based forwarding

The simplest version of geographic forwarding, described by Finn [53], is to forward the packet at each step to the neighbor that is closest to the destination (the location or ID of this neighbor is marked in the packet, based on neighborhood information obtained through a beaconing process). There are several variants of this greedy forwarding mechanism. In the most forward within R strategy (MFR) [208], the packet is forwarded to the neighbor whose projection on the line joining the current node and the destination is the farthest (note that this is not always the same as the greedy forwarding). In another variant, the packet is forwarded to the nearest neighbor with forward progress (NFP) [85], so that contention can be minimized and the wireless link quality is high. Yet another variant is the random-forwarding technique in which the packet is forwarded at random to a neighbor so long as it makes forward progress. We should also note that the ExOR technique, described above, is a form of implicit greedy geographic forwarding, which does not require beaconing – the timers are designed so that packets are likely to be forwarded (only) by the node closest to the destination that receives the packet correctly.

8.6.2 Perimeter routing (GFG/GPSR)

One major shortcoming of the greedy forwarding technique is that it is possible for it to get stuck in local maxima/dead-ends. Such dead-ends occur (see Figure 8.5) when a node has no neighbors that are closer to the destination than itself. For this case, the greedy-face-greedy (GFG) algorithm [14], which is the basis of the greedy perimeter stateless routing protocol (GPSR) [102], routes along the face of a planar sub-graph using the right-hand rule. The planar sub-graph can be obtained using localized constructions such as the Gabriel graph and the relative neighbor graph constructions. A packet switches from greedy to face-routing mode whenever it reaches a dead end, is then routed using face routing, and then reverts back to greedy mode as soon as it reaches a node that is closer to the destination than the dead-end node. Other studies have examined ways to improve upon and get provable efficiency guarantees with face-routing approaches. It should be kept in mind, however, that the likelihood that such dead ends exist decreases with network density; it can be shown that, if the graph is dense enough that each interior node has a neighbor in every $2\pi/3$ angular sector, then greedy forwarding will always succeed [49].

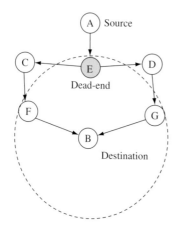

Figure 8.5 A dead-end for greedy geographic forwarding

8.6.3 The PRR*D metric

Seada *et al.* show in [187] that greedy geographic forwarding techniques exhibit a distance-hop–energy tradeoff over real wireless links. If at each step the packet travels a long distance (as in the basic greedy and MFR techniques), the total number of hops is minimized; however, because of the distance, each hop is likely to be a weak link, with poor reception rate requiring multiple transmissions for successful delivery. At the other extreme, if the packet travels only a short distance at each hop (as with the NFP technique), the links are likely to be good quality; however there are multiple hops to traverse, which would increase the number of transmissions as well. By extensive simulations, real experiments as well as analysis, it is shown that the localized metric that maximizes the energy efficiency while providing a good end-to-end delivery reliability is the product of the packet reception rate on the link (probability of successful delivery) and the distance improvement towards the destination D, which is known as the $PRR * D$ metric. Thus, when it has a packet to transmit to a given destination, each node selects the neighbor with the highest $PRR * D$ metric to be the one to forward the message further.

8.6.4 Geographical energy-aware routing (GEAR)

The geographical and energy-aware routing mechanism (GEAR) [240] is designed to propagate queries/interests from a sink to all sensor nodes in a geographically scoped destination region (e.g. a square sub-area with specified bounds). The packets are routed from the origin to the destined region through a series of forwarding steps, which make use of an adaptive learned cost function for each neighbor that is modified over time to provide a balance between reach-ability and energy efficiency. This approach also provides robustness to dead ends. When the packet reaches the destination region, a recursive forwarding technique is employed as follows. The region is split into k sub-regions, and a packet is forwarded to each sub-region. This split–forward sequence is repeated until the region contains only one node, at which point the packet has been successfully propagated to all nodes within the query region.

Finally, we should mention in this section the trajectory-based forwarding technique (TBF) [148], which is also an important geographic routing technique for sensor networks. As a significant application of TBF applies to the routing of queries, however, we shall defer its description to the next chapter.

8.7 Routing to mobile sinks

Mobility of nodes in the network adds a significant challenge. The study of routing over mobile ad hoc networks (MANET) has indeed been an entire field in itself, with many protocols such as DSR, AODV, ZRP, ABR, TORA, etc. proposed to provide robustness in the face of changing topologies [158, 211]. A thorough treatment of networking between arbitrary end-to-end hosts in the case where all nodes are mobile is beyond the scope of this text. However, even in predominantly static sensor networks, it is possible to have a few mobile nodes. One scenario in particular that has received attention, is that of *mobile sinks*. In a sensor network with a mobile sink (e.g. controlled robots or humans/vehicles with gateway devices), the data must be routed from the static sensor sources to the moving entity, which may not necessarily have a predictable/deterministic trajectory. A key advantage of incorporating mobile sinks into the design of a sensor network is that it may enable the gathering of timely information from remote deployments, and may also potentially improve energy efficiency.

We describe below a few studies that propose solutions to different variants of this problem.

8.7.1 Two-tier data dissemination (TTDD)

In the two-tier data dissemination (TTDD) approach [239], all nodes in the network are static, except for the sinks that are assumed to be mobile with unknown/uncontrolled mobility. The data about each event are assumed to originate from a single source. Each active source creates a grid overlay dissemination network over the static network, with grid points acting as dissemination nodes (see Figure 8.6). A mobile sink, when it issues queries for information, sends out a locally controlled flood that discovers its nearest dissemination point. The query is then routed to the source through the overlay network. The sink includes in the query packet information about its nearest static neighbor, which acts as a *primary agent*. An alternative *immediate agent* is also chosen when the sink is about to go out of reach of the primary agent for robust delivery. The source sends data to the sink through the overlay dissemination network to its closest grid dissemination node, which then forwards it to its primary agent. As the sink moves through the network, new primary agents are selected and the old ones time out; when a sink moves out of reach of its nearest dissemination node, a new dissemination node is discovered and the process continues.

8.7.2 Asynchronous dissemination to mobile sinks (SEAD)

The scalable energy-efficient asynchronous dissemination technique (SEAD) presented in [107] provides for communication of information from a given source in a static sensor network to multiple mobile sinks. Each mobile sink selects a nearby static access node to communicate information to and from the source. Only the access node keeps track of sink movement, so long as it does not move too far away. When the hop-count between the sink and the nearest access point exceeds a threshold, a new access node is selected by the sink. Data are sent from the source first to the various access nodes through a dynamically

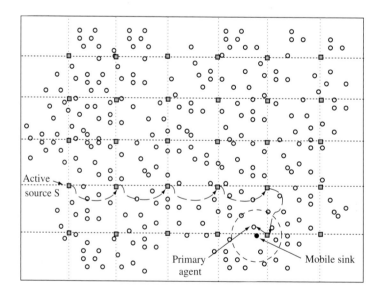

Figure 8.6 The two-tier data dissemination technique

maintained multi-cast dissemination tree, which then forwards the information to their respective sinks. The multi-cast tree construction scheme described for SEAD relies on the replication of source data at multiple points in the network. Each access node selects a nearby replica to obtain data from, based on a minimum-cost increase criterion.

8.7.3 Data mules

For sparsely deployed sensor networks (e.g. deployed over large areas), the network may never be truly connected; in the most extreme case no two sensor devices may be within radio range of each other. The MULE (mobile ubiquitous LAN extensions) architecture [190] aims to provide connectivity in this environment through the use of mobile nodes that may help transfer data between sensors and static access points, or may themselves act as mobile sinks.

It is assumed that MULE nodes do not have controlled mobility and that their movements are random and unpredictable. Whenever a MULE node comes into contact with a sensor node, it is assumed that all the sensors data are transferred over to the MULE. Whenever the MULE comes into contact with an access point, it transfers all information to the access point. It is assumed that there is sufficient density of MULE nodes, with sufficiently uniform mobility, so that all sensor nodes can be served, although delays are likely to be quite high. Both MULEs and sensors are assumed to have limited buffers, so that new data are stored or transferred only if there is buffer space available.

The MULE architecture has been analyzed using random walks on a grid [190]. The analysis provides an insight into the scaling behavior of this system with respect to number of sensors, MULEs, and access points. One conclusion of the study worth noting is that the buffer size of MULE nodes can be increased to compensate for fewer MULE nodes in terms of delivery rates (albeit at the expense of increased latency), but increasing sensor buffers alone does not necessarily have a similar effect.

8.7.4 Learning enforced time domain routing

The hybrid learning enforced time domain routing (HLETDR) technique (Figure 8.7) is proposed for situations where the sink follows a somewhat predictable but stochastic repeated pattern [7]. It is assumed that events happen rarely, so that the sink does not issue queries, but rather that sources need to push data about events towards the sink. Nodes near the path of the sink's movements, called moles, *learn* the probability distribution of the sink's appearance in their

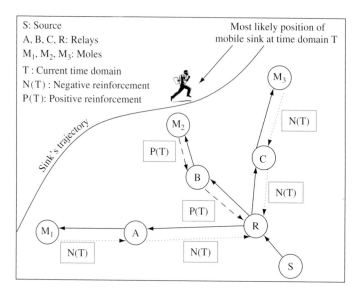

Figure 8.7 Illustration of hybrid learning enforced time domain routing

vicinity. The periodic time between tours of the sink is divided into multiple domains, such that the sink may be more likely to be in the vicinity of one set of moles in one time domain, and in the vicinity of another set of moles in another time domain.

For each time domain, local forwarding probabilities are maintained at intermediate nodes. When data are generated, depending on the time, they are routed through the intermediate nodes based on these probabilities to try and reach a mole that the sink will pass by. Initially, the probability weights at nodes are all equal, resulting in unbiased random walks. Over time, these weights are reinforced positively or negatively by moles, depending on the sink probability distribution and success of the data delivery. Multiplicative weight update rules for reinforcements are found to be most efficient and robust. A few iterations may suffice to determine efficient routes for data to reach a mole that is highly likely to encounter the sink and be delivered successfully.

8.8 Summary

We have examined a number of issues and design concepts relevant to reliable, energy-efficient routing in wireless sensor networks: selection of routing metrics, multi-path routing, geographic routing, and delivery of data to mobile nodes.

Reliability in routing can be achieved in many ways. One approach is metric based. The ETX metric requires nodes in the network to monitor the quality of links to their neighbors to find routes that minimize the number of transmissions incurred due to ARQ retries. The MOR/MER metrics, developed analytically, are better suited for rapidly time-varying channels and offer an alternative way to effect energy–reliability tradeoffs. Another approach to reliable data delivery is the use of multi-path routing techniques, such as the braided multi-path mechanism, and the GRAd, GRAB routing protocols. Finally, reliability can be enhanced even at the network layer by exploiting the spatial diversity of independent fading channels, as in the relay diversity technique and the extremely opportunistic routing mechanism.

Energy-aware routing techniques utilize metrics that take into account the residual lifetimes of intermediate nodes on the routing path. They can be enhanced with probabilistic forwarding to provide some degree of energy–load balancing. Global solutions in the form of flow-based optimization formulations have also been used to analyze lifetime-maximizing flows and develop distributed algorithms for energy-efficient routing.

Given that nodes in a sensor network are likely to have at least coarse-grained position information, geographic forwarding techniques, which provide low-overhead stateless routing, can be an attractive option.

Finally, while our focus has largely been on sensor networks with static nodes, in some applications it may be necessary to include limited mobility in the form of a moving sink. Routing techniques such as TTDD, SEAD, and HLETDR address this scenario.

Exercises

8.1 *ETX:* For the directed graph labelled with reception probabilities shown in Figure 8.3 (ignore the priorities associated with ExOR, and assume $d_r = 1$ on all links), determine the optimal ETX route from node A to node B.

8.2 *Relay diversity and MAC:* Explain why the relay diversity scheme may not work well with some sleep-oriented MAC protocols proposed for sensor networks.

8.3 *Relay diversity:* For the example of relay diversity shown in Figure 8.2, say the probabilities of reception for the links A–B, B–C, and A–C were 0.8, 0.8, and 0.6 respectively. What is the probability of successful reception at C without and with relay diversity?

8.4 *Timer-based ExOR routing:* Consider a variant of ExOR routing in which successful recipients of a message set a timer for retransmission of the message depending upon their priority. Thus, in the example of Figure 8.3, node B's timer would be set to 1, node D to 2, node E to 3, node C to 4. If a node hears another one forwarding the message, it cancels its timer. After a node's timer expires, it will forward the message itself. In this case, once A sends the packet, what is the expected delay before the first of its recipients forwards the message?

8.5 *Flow formulation:* Adapt the linear program given in Section 8.5.4 for a fairness-oriented objective function that maximizes the minimum flow rate from all sources.

8.6 Greedy geographic routing fails when a forwarding node on the path finds no neighbors within range that are closer than itself to the destination. Prove that this implies the existence of a $2\pi/3$ angular sector centered at this node in which it has no neighbors.

8.7 *MULE simulation study:* On a 10×10 square grid, place the sink node at the bottom-left-most grid, and ten sources at random squares. Simulate the movement of k MULE nodes (with varying k), such that all execute independent unbiased random walks on the grid, moving to a neighboring cell at each time step. Assume the MULE nodes pick up one unit of information from the source when they are in the same grid, and drop off all information they contain when they arrive at the sink. Assume that the sources always have data available for pick-up and an infinite buffer, and that MULEs have an infinite buffer too. Analyze significant metrics, such as the average time delay between each visit to the sink, average size of the MULE buffers, average throughput between sources and the sink, as functions of the number of MULE nodes. What is the impact of placing additional static sink nodes?

Data-centric networking

9.1 Overview

A fundamental innovation in the area of wireless sensor networks has been the concept of *data-centric networking*. In a nutshell, the idea is this: routing, storage, and querying techniques for sensor networks can all be made more efficient if communication is based directly on application-specific data content instead of the traditional IP-style addressing [74].

Consider the World Wide Web. When one searches for information on a popular search site, it is possible to enter a query directly for the content of interest, find a hit, and then click to view that content. While this process is quite fast, it does involve several levels of indirection and names: ranging from high-level names, like the query string itself, to domain names, to IP (internet protocol) addresses, and MAC addresses. The routing mechanism that supports the whole search process is based on the hierarchical IP addressing scheme, and does not directly take into account the content that is being requested. This is advantageous because the IP is designed to support a huge range of applications, not just web searching. This comes with increased indirection overhead in the form of the communication and processing necessary for binding; for instance the search engine must go through the index to return web page location names as the response to the query string, and the domain names must be translated to IP addresses through DNS. This tradeoff is still quite acceptable, since the Internet is not resource constrained.

Wireless sensor networks, however, are qualitatively different. They are application specific so that the data content that can be provided by the sensors is relatively well defined *a priori*. It is therefore possible to implement network operations (which are all restricted to querying and transport of raw and processed sensor data and events) directly in terms of named content. This data-centric approach to networking has two great advantages in terms of efficiency:

1. Communication overhead for binding, which could cause significant energy wastage, is minimized.
2. In-network processing is enabled because the content moving through the network is identifiable by intermediate nodes. This allows further energy savings through data aggregation and compression.

9.2 Data-centric routing

9.2.1 Directed diffusion

One of the first proposed *event-based* data-centric routing protocols for WSN is the directed diffusion technique (Figure 9.1) [96, 97].

This protocol uses simple attribute-based naming as the fundamental building block. Both requests for information (called *interests*) and the notifications of observed events are described through sets of attribute–value pairs. Thus, a request for 10 seconds worth of data from temperature sensors within a particular rectangular region may be expressed as follows:

```
type     = temperature      // type of sensor data
start    = 01:00:00         // starting time
interval = 1s               // once every second
duration = 10s              // for ten seconds
location = [24, 48, 36, 40] // within this region
```

And one of the data responses from a particular node may be:

```
type      = temperature // type of sensor data
value     = 38.3         // temperature reading
timestamp = 01:02:00     // time stamp
location  = [30, 38]     // x,y coordinate
```

The steps of the basic directed diffusion are as follows:

1. The sink issues an interest for a specific type of information that is flooded throughout the network (the overhead of this can be reduced if necessary by using geographic scoping or some other optimization). The interest may be periodically repeated if robustness is called for.
2. Every node in the network caches the interest while it is valid, and creates a local gradient entry towards the neighboring node(s) from which it heard

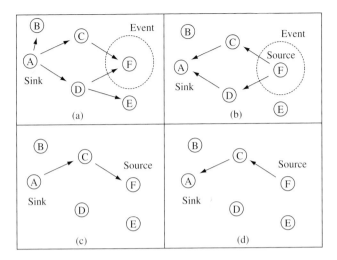

Figure 9.1 Directed diffusion routing

the interest. The sink's ID/network address is not available and hence not recorded, however the local neighbors are assumed to be uniquely identifiable through some link-layer address. The gradient also specifies a value (which could be an event rate, for instance).

3. A node which obtains sensor data that matches the interest begins sending its data to all neighbors it has gradients toward. If the gradient values stand for event rates then the rate to each neighbor must satisfy the gradients on the respective link. All received data are cached in intermediate nodes to prevent routing loops.

4. Once the sink starts receiving response data to its interest from multiple neighbors, it begins reinforcing one particular neighbor (or k neighbors, in case multi-path routing is desired), requesting it to increase the gradient value (event rate). These reinforcements are propagated hop by hop back to the source. The determination of which neighbor to reinforce can take into account other considerations such as delay, link quality, etc. Nodes continue to send data along the outgoing gradients, depending on their values.

5. (Optional) Negative reinforcements are used for adaptability. If a reinforced link is no longer useful/efficient, then negative reinforcements are sent to reduce the gradient (rate) on that link. The negative reinforcements could be implemented by timing out existing gradients, or by re-sending interests with a lower gradient value.

Essentially what directed diffusion does is (a) the sink lets all nodes in the network know what the sink is looking for, (b) those with corresponding data respond by sending their information through multiple paths, and (c) these are pruned via reinforcement so that an efficient routing path is obtained.

The directed diffusion mechanism presented here is highly versatile. It can be extended easily to provide multi-path routing (by changing the number of reinforced neighbors) as well as routing with multiple sinks/sources. It also allows for data aggregation, as the data arriving at any intermediate node from multiple sources can be processed/combined together if they correspond to the same interest.

9.2.2 Pull versus push diffusion

The basic version of directed diffusion described above can be viewed as a *two-phase pull* mechanism. In phase 1, the sink *pulls* for information from sources with relevant information by propagating the interest, and sources respond along multiple paths; in phase 2, the sink initiates reinforcement, then sources continue data transfer over the reinforced path. Other variants of directed diffusion include the *one-phase pull* and the *push diffusion* mechanisms [75].

The two-phase pull diffusion can be simplified to a one-phase pull mechanism by eliminating the reinforcements as a separate phase. In one-phase pull diffusion, the sink propagates the interest along multiple paths, and the matching source directly picks the best of its gradient links to send data and so on up the reverse path back to the sink. While potentially more efficient than two phase pull, this reverse-path selection assumes some form of bidirectionality in the links, or sufficient knowledge of the link qualities/properties in each direction.

In push diffusion, the sink does not issue its interests. Instead sources with event detections send exploratory data through the network along multiple paths. The sink, if it has a corresponding interest, reinforces one of these paths and the data-forwarding path is thus established.

The push and pull variants of diffusion have been compared and analyzed empirically [75] as well as through a simple mathematical model [111]. The results quantify the intuition that the pull and push variants are each appropriate for different kinds of applications. In terms of the route setup overhead, pull diffusion is more energy-efficient than push diffusion whenever there are many sources that are highly active generating data but there are few, infrequently interested sinks; while push diffusion is more energy-efficient whenever there are few infrequently active sources but there are many frequently interested sinks.

The threshold-sensitive energy-efficient sensor network protocol (TEEN) [132] is another example of a push-based data-centric protocol. In TEEN, nodes react immediately to drastic changes in the value of a sensed attribute and when this change exceeds a given threshold communicate their value to a cluster-head for forwarding to the sink.

9.3 Data-gathering with compression

Several researchers have investigated the combination of gathering information in a WSN by combining routing with in-network compression. While the exact type of compression involved can be quite application specific, these studies reveal a number of general principles and the tradeoffs involved. In most of these studies, the efficiency metric of interest is the total number of data bit transmissions (i.e. cumulative number of bits over each hop of transmission) per round of data-gathering from all sources. Besides providing energy efficiency by reducing the amount of transmissions, combining routing with compression also has the potential to improve network data throughput in the face of bandwidth constraints [186]. We now describe some of the pertinent techniques and analytical studies.

9.3.1 LEACH

The LEACH protocol [76] is a simple routing mechanism proposed for continuous data-gathering applications. In LEACH, illustrated in Figure 9.2, the

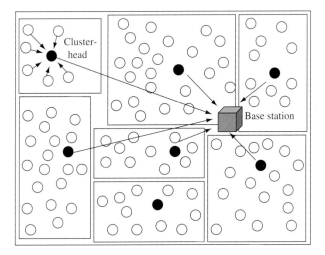

Figure 9.2 The LEACH cluster-based routing technique

network is organized into clusters. The cluster-heads periodically collect and aggregate/compress the data from nodes within the cluster using TDMA, before sending them to the sink. The cluster-heads may send to the sink through a direct transmission or through multiple hops. Cluster-heads are rotated periodically for load balancing.

9.3.2 Distributed source coding

There exist distributed source coding techniques based on the Slepian–Wolf theorem [34] that allow joint coding of correlated data from multiple sources without explicit communication, so long as the individual source rates satisfy certain constraints pertaining to different conditional entropies. These techniques require that the correlation structure be available in advance at the independent coders. While this approach increases the complexity of the coding and requires upfront collection of information about joint statistics (which may not be feasible in all practical conditions), it effectively makes routing and coding decisions independent of each other, since the independently coded data can be sent along the shortest paths to the sink. In a networked context, then, the only design consideration for energy efficiency is to ensure that sources near the destination are allocated higher rates [35].

9.3.3 Impact of compression

The gains due to in-network compression can be best demonstrated in the extreme case where the data from any number of sources can be combined into a single packet (e.g. duplicate suppression, when the sources generate identical data). In this case, if there are k sources, all located close to each other and far from the sink, then a route that combines their information close to the sources can achieve a k-fold reduction in transmissions, as compared with each node sending its information separately without compression. In general, the optimal joint routing–compression structure for this case is a minimum Steiner tree construction problem, which is known to be NP-hard. However there exist polynomial solutions for special cases where the sources are close to each other [109].

9.3.4 Network correlated data-gathering

Cristescu, Beferull–Lozano, and Vetterli [35] consider the case where all nodes are sources but the level of correlation can vary. In this case, when the data are completely uncorrelated then the shortest path tree provides the best solution (in minimizing the total transmission cost). However, the general case is treated by choosing a particular correlation model that preserves the complexity; in this model, only nodes at the leaf of the tree need to provide R bits, but all other interior nodes, which have side information from other nodes, need only generate r bits of additional information. The quantity $\rho = 1 - \frac{r}{R}$ is referred to as the correlation coefficient. Now, it can be shown that a travelling salesman path (that visits all nodes exactly once) provides an arbitrarily more efficient solution compared with shortest path trees as ρ increases. It is shown that this problem is NP-hard for arbitrary ρ vlaues.

A good approximation solution for the problem is the following combination of SPT and the travelling salesman paths. All nodes within some range of the sink (larger the ρ, smaller this range) are connected through shortest path trees, and beyond that each strand of the SPT is successively grown by adding nearby nodes, an approximate way to construct the travelling salesman paths. Thus the data from distant nodes are compressed sequentially up to a point, and then sent to the sink using shortest paths.

9.3.5 Simultaneous optimization for concave costs

Goel and Estrin [64] treat the case when the exact reduction in data that can be obtained by compressing k sources is not known. The only assumption that is made is that the amount of compression is concave with respect to k. This is a very reasonable assumption, as it essentially establishes a notion of monotonically diminishing contributions to the total non-redundant information; it simply means that the additional non-redundant information in the $j + 1$th source is smaller than that of the jth source. A random tree construction is developed for this problem that ensures that the expected cost is within a factor $O(\log k)$ of the optimal, regardless of what the exact concave compression function is.

9.3.6 Scale-free aggregation

In practice the degree of spatial correlation is a function of distance, and better approximations are possible by taking this into account. Nodes nearby are able to provide higher compression than nodes at a greater distance. A model for the spatial correlations that captures this notion is considered [45]. Each node in a square grid of sensors is assumed to have information about the readings of all nodes within a k-hop radius. Nodes can communicate with any of their four cardinal neighbors. The aggregation/compression function considered is such that any redundant readings are suppressed in the intermediate hops. This work assumes a square grid network in which the source is located at the origin, on the bottom-left corner. The routing technique proposed is a randomized one: a node at location (x, y) forwards its data, after combining with any other preceding sources sending data through it, with probability $x/(x+y)$ to its left neighbor and with probability $y/x+y$ to its bottom neighbor. It is shown that this randomized routing technique can provide a constant factor approximation in expectation to the optimal solution.

9.3.7 Impact of spatial correlations on routing with compression

In [153], an empirically derived approximation is used to quantify spatial correlation in terms of joint entropies. The total joint information generated by an arbitrary set of nodes is obtained using an approximate incremental construction. At each step of this construction, the next nearest node that is at a minimum distance d_{min} to the current set of nodes is considered. This node contributes an amount of uncorrelated data equal to $\frac{d_{min}}{c+d_{min}} \cdot H_1$, where H_1 is the entropy of a single source and c a constant that characterizes the degree of spatial correlation. In the simplest case when all nodes are located on a line with equal spacing of d, this procedure yields the following expression for the joint entropy of n nodes:

$$H_n = H_1 + (n-1)\frac{d}{c+d} \cdot H_1 \tag{9.1}$$

Consider first the two extremes: (i) when $c = 0$, $H_n = nH_1$, the nodes are completely uncorrelated; (ii) when $c \to \infty$, $H_n = H_1$, the nodes are completely correlated.

Under this model, it becomes easy to quantify the total transmission cost of any tree structure where routing is combined with en route compression. An example scenario is considered with a linear set of sources at one end of a square grid communicating their data to a sink at the other end. An idealized distributed source coding is used as a lower bound for the communication costs in this setting. It is shown that at one extreme, when the data are completely uncorrelated ($c = 0$), the best solution is that of shortest path routing (since there is no possible benefit due to compression). At the other extreme, when data are perfectly correlated ($c \to \infty$), the best solution is that of routing the data among the sources first so that they are all compressed, before sending the combined information directly to the sink. For in-between scenarios, a clustering strategy is advocated such that the data from s nearby sources are first compressed together, then routed to the sink along the shortest path. It is shown that there is an optimal cluster size corresponding to each value of the correlation parameter. The higher the level of correlation, the larger the optimal cluster size. However, surprisingly, it is also found that there exists a near-optimal cluster size that depends on the topology and sink placement but is insensitive to the exact correlation level.

This result has a practical implication, because it suggests that a LEACH-like clustering strategy combined with compression at the cluster-heads can provide an efficient solution even in the case of heterogeneous or time-varying correlations.

9.3.8 Prediction-based compression

Another approach to combining routing and compression is to perform prediction-based monitoring [63]. The essence of this idea is that the base station (or a cluster-head for a region of the network) periodically gathers data from all nodes in the network, and uses them to make a prediction for data to be generated until the next period. In the simplest case, the prediction may simply be that the data do not change. More sophisticated predictions may indicate how the data will change over time (e.g. the predictions may be based on the expected movement trajectory of a target node, or in the case of diffuse phenomena such as heat and chemical plumes, these predictions may even be based on partial differential equations with known or estimated parameters [176]). This prediction is then broadcast to all nodes within the region. During the rest of the period, the component nodes only transmit information to the base station if their measurements differ from the predicted measurements.

9.3.9 Distributed regression

A closely related technique is the use of a distributed regression framework for model-based compression [69]. In this approach, nodes collaboratively determine the parameters of a globally parameterized function using local measurements. The model used is a weighted sum of local basis functions (which could just be polynomials, for example). Distributed kernel linear regression techniques are then used to determine the parameters. Compression is achieved by transmitting only the model parameters instead of the full data.

9.4 Querying

In basic data-gathering scenarios, such as those discussed above in connection with compression, information from all nodes needs to be provided continuously to the sink. In many other settings, the sink may not be interested in all the information that is sensed within the network. In such cases, the nodes may store the sensed information locally and only transmit it in response to a query issued by the sink. Therefore the querying of sensors for desired information is a fundamental networking operation in WSN. Queries can be classified in many ways:

1. **Continuous versus one-shot queries**: depending on whether the queries are requesting a long duration flow or a single datum.
2. **Simple versus complex queries**: complex queries are combinations that consist of multiple simple sub-queries (e.g. queries for a single attribute type); e.g. "What are the location and temperature readings in those nodes in the network where (a) the light intensity is at least w and the humidity level is between x and y OR (b) the light intensity is at least z." Complex queries may also be aggregate queries that require the aggregation of information from several sources; e.g. "report the average temperature reading from all nodes in region R1."
3. **Queries for replicated versus queries for unique data**: depending on whether the queries can be satisfied at multiple nodes in the network or only at one such node.
4. **Queries for historic versus current/future data**: depending on whether the data being queried for were obtained in the past and stored (either locally at the same node or elsewhere in the network), or whether the query is for current/future data. In the latter case data do not need to be retrieved from storage.

When the queries are for truly long-term continuous flows, the cost of the initial querying may be relatively insignificant, even if that takes place through naive flooding (as for instance, with the basic directed diffusion). However, when they are for one-shot data, the costs and overheads of flooding can be prohibitively expensive. Similarly, if the queries are for replicated data, a flooding may return multiple responses when only one is necessary. Thus other alternatives to flooding-based queries (FBQ) are clearly desirable.

9.4.1 Expanding ring search

One option is the use of an expanding ring search, illustrated in Figure 9.3. An expanding ring search proceeds as a sequence of controlled floods, with the radius of the flood (i.e. the maximum hop-count of the flooded packet) increasing at each step if the query has not been resolved at the previous step. The choice of the number of hops to search at each step is a design parameter that can be optimized to minimize the expected search cost using a dynamic programming technique [22].

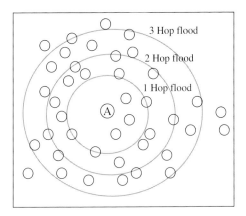

Figure 9.3 Expanding ring search

However, in the absence of replicated/cached information in the network, for arbitrarily located data, expanding ring searches do not provide useful gains over flooding. This is demonstrated in [24], which shows that less than 10% energy savings are obtained from expanding ring search compared with flooding when there is no caching/replication of data in the network, while the delay increases significantly. Intuitively, when there is replicated information, expanding ring searches are more likely to offer significant improvements, because the likelihood of resolving the query earlier would be higher. However, in this situation, other approaches may potentially provide better improvements.

9.4.2 Information-driven sensor query routing (IDSQ)

The information-driven sensor querying approach (IDSQ) [28] suggests an incremental approach to sensor tasking that is suitable for resource constrained, dynamic environments. The problem of how to route the query to a node with the maximum information gain is a core problem, that is addressed by the constrained anisotropic diffusion routing (CADR) technique. CADR essentially routes the query through a greedy search, making a sequence of local decisions at intermediate steps, based on sensor values of neighboring nodes. A composite objective function that combines the information utility and communication costs is first defined. These decisions can be made in different ways:

- by forwarding the query to the neighboring node with the highest objective function;
- by forwarding the query to the neighboring node with the steepest (local) gradient in the objective function;

- by forwarding the query to the neighboring node which maximizes a combination of the local gradient of the objective function and the distance improvement to the estimated optimum location (the information gain formulation used in IDSQ allows an estimation of the geographic location of the query destination).

One advantage of this approach, which provides a greedy descent towards the query destination, is that, if partial solutions are shipped back to the query-originating node, it is provided with incrementally better information as the query moves towards the global optimum. Further, depending on how the objective function is designed, this technique can minimize the energy needed to route the query to the destination by choosing the shortest path, or it can maximize the information gain by taking an irregular walk with more steps.

A related work on multi-step information-driven querying [122] aims to provide a minimum hop path through the network that maximizes the accumulated information gain. The following approach provides a solution: for each node i assign all links going into that node a cost of $L - u_i$, where L is a sufficiently large number and u_i the information utility at node i. Then find a shortest path from the origin to destination using this link metric (e.g., using Dijkstra's algorithm). This choice of cost function minimizes the number of hops, because of the large additive L terms. Among the minimum hop paths, the algorithm also maximizes the accumulated utility, because of the $-u_i$ terms in the minimization expression.

9.4.3 Fingerprint gradients (RUGGED)

The technique of query routing using fingerprint gradients (RUGGED) proposed by Faruque and Helmy [51] also makes use of sensor readings within the network to send the query to the node with the highest reading, which is assumed to be the node closest to an event source. RUGGED switches forwarding modes depend on the information available: if no gradient information is available in a region (i.e. far away from phenomena), then flooding is utilized; in the gradient information region, a greedy forwarding approach is utilized whenever distance improvement is possible, or else probabilistic forwarding is used to escape local minima. The parameters of the probabilistic forwarding can be varied depending on the sensor readings.

9.4.4 Trajectory-based forwarding (TBF)

In WSN where nodes in the network all have reasonably accurate location information (either directly through GPS or through the implementation of a network localization technique), a unique approach to efficient querying is the use of pre-programmed paths embedded into the query packet. The geographic trajectory-based forwarding (TBF) technique [148] provides this functionality.

The source encodes a trajectory for the query packet into the header. The trajectory could be anything that can be represented in a parametric form $(x(t), y(t))$ (though non-parametric representations are also possible in principle). For instance a packet to be sent along a sinusoidal curve in a single direction would have the trajectory encoding $(x(t) = t, y(t) = A \sin t)$; and, to travel on a straight line with slope α, it would have the encoding $(x(t) = t \cos \alpha, y(t) = t \sin \alpha)$.

During the course of the forwarding, the ith node that receives the packet with the encoded trajectory determines the corresponding time t_i as the value of t that corresponds to the point of the curve closest to its location (if the curve passes by the same location more than once, then additional information, such as the parameter value chosen by the previous node in the forwarding, may be utilized to determine t_i). This node then examines its neighboring nodes to determine which of them would be most suitable to forward the packet to, depending on $x(t)$, $y(t)$, and t_i. To make progress on the trajectory, the next hop neighbor must have a parameter value t_{i+1} higher than t_i.

The next hop can be determined in many ways depending on design considerations, such as by (a) picking the neighbor offering the maximum distance improvement, (b) picking the neighbor that offers the minimum deviation from the encoded trajectory, (c) picking the node closest to the centroid of the candidate neighbors, and (d) picking the node with maximum energy. Repeating this process at each step, the packet will follow a trajectory close to that specified by the parametric expression in the packet. This is illustrated in Figure 9.4. Good features of this technique are that the trajectory information can often be represented quite compactly, a number of different types of trajectories can be encoded, and the forwarding decisions at each step are local and dynamic. The denser the network, the more accurately will the actual trajectory match the desired trajectory.

While it has many possible applications, TBF is uniquely suited for propagating queries within the network. When a set of possible locations must all be visited, TBF provides an efficient way to guide the query.

9.4.5 Active query forwarding (ACQUIRE)

ACQUIRE [177, 178] is a querying technique involving active query forwarding. The idea is to treat the query as an intelligent entity that moves through the network searching for the desired response. If the query is a complex query, its component sub-queries can be resolved partially en route. As shown in Figure 9.5,

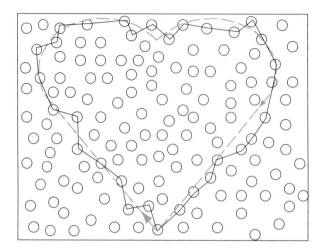

Figure 9.4 Trajectory-based forwarding

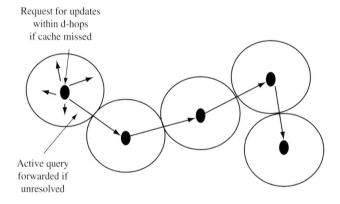

Figure 9.5 ACQUIRE

ACQUIRE progresses through the network as a repeated sequence of three parts:

1. Examine cache: When the query arrives at a node in the network (referred to as an active node), the node first checks its existing cache to see if its cache is valid/fresh enough to directly resolve the query (or parts of it). If the entire query can be resolved from the active node's cache, the response is returned to the sink, else the process continues as follows.

2. Request updates: If the cache does not contain the information desired, the active node issues a request for updates from nodes within a d-hop neighborhood. The responses from the controlled flood are then gathered back and used to see if the query can be resolved.

3. Forward: If it has not already been resolved, the query is then forwarded to another active node (chosen either randomly or through some guided mechanism such as TBF) by a sufficient number of hops so that the controlled flood phases (described below) do not overlap significantly.

A key observation about ACQUIRE is that the look-ahead parameter offers a tunable tradeoff between a trajectory-based query when $d = 0$ (which could be either a random walk or a guided trajectory, depending on the implementation) and a full flood when $d = D$, the diameter of the network. There is a tradeoff for different values of the look-ahead parameter d; when the value of d is small, the query needs to be forwarded more often, but there are fewer update messages at each step. When d is large, fewer forwarding steps are involved, but there are more update messages at each step.

The optimal choice of d in ACQUIRE depends most on sensor data dynamics, which can be captured by the ratio of the rate at which data change in the network to the rate at which queries are generated. When the data dynamics are low, caches remain valid for a long time and therefore the cost of a large d flood can be amortized over several queries; however, when the data dynamics are very high, repeated flooding is required, and hence a small d is favored.

9.4.6 Rumor routing

The rumor routing technique [15] provides an efficient rendezvous mechanism to combine push and pull approaches to obtain the desired information from the network. In rumor routing, sinks desiring information send queries through the network, while sources generating important events send notifications through the network. These are both treated as mobile agents. The event notifications leave a "sticky" trail of state information through the network. Then, a query agent visiting a node where an event notification agent has already passed will find pointer information on the location of the corresponding source. This is shown in Figure 9.6.

Thus, it suffices for the queries to simply intersect with one of the event notification trajectories, rather than have to locate the event node itself. An additional optimization provided is the ability of event notification agents to propagate information about other events based on the state encountered in intermediate nodes, which can reduce the number of unique agents generated for each event.

The trajectory followed by both the events and the queries can be either a random walk (with some loop-prevention built in), or more directed, e.g. straight lines generated using a TBF scheme. It is shown that substantial improvements

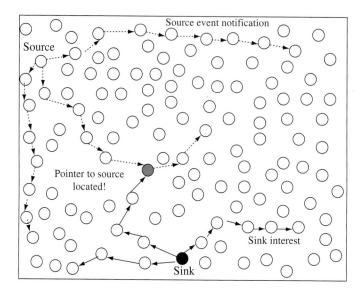

Figure 9.6 Rumor routing

in the energy costs can be obtained by rumor routing compared with the two extremes of query flooding (pull) and event flooding (push).

9.4.7 The comb-needle technique

The same approach as followed by rumor routing, of combining push and pull by looking at intersections of queries and event notifications, is also the basis of the comb-needle technique [124].

In the basic version of this technique, illustrated in Figure 9.7 the queries build a horizontal comb-like routing structure, while events follow a vertical needle-like trajectory to meet the "teeth" of the comb. A key tunable parameter in this construction is spacing between branches of the comb and correspondingly the length of the trajectory traversed by event notifications, which can be adjusted depending on the frequency of queries as well as the frequency of events. To minimize the average total cost per query, the comb inter-spacing as well as the length of the event trajectories should be smaller when the event-to-query ratio is higher (more pull, less push); however, when there the event-to-query ratio is lower, the comb inter-spacing as well as the distance traversed by even notifications should be higher (less pull, more push).

In practice, the frequency of both queries and events is likely to fluctuate over time. An adaptive version of the algorithm [124] handles this scenario. In this adaptive technique, the inter-comb spacing and needle trajectory length are

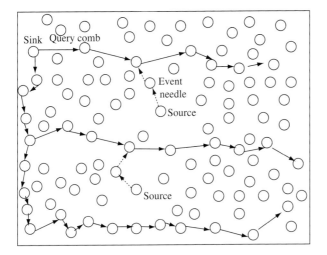

Figure 9.7 The comb–needle approach

calculated and adjusted dynamically in a distributed manner by the sources and sinks by locally estimating data and query frequencies over several observations.

The basic comb structure, with the queries forming the horizontal comb and events the vertical needles is a *global-pull, local-push* model, best suited for conditions when the frequency of queries is less than the frequency of events. When the frequency of queries is more than the frequency of events, and there are multiple querying nodes, then a *reverse-comb structure* can be employed, which would provide a *global-push, local-pull* structure. In the reverse-comb structure, the events form the vertical comb, while queries take horizontal needle trajectories to match with events.

9.4.8 Asymptotics of query strategies

An analytical study of various query strategies by Shakkottai [192] considers three types of queries, all executed using random walks that time out on average after t time units: (i) the source-driven query, in which a random walk from the source attempts to reach within some ϵ region near the location of the desired information, (ii) a replication strategy, where the information is replicated regularly throughout the network, but the search is still source driven, and (iii) a source–receiver "sticky" search, in which, as in the rumor routing and the comb–needle strategies, both the source and destination initiate k random walks and it suffices for any source-initiated walk to intersect with any receiver-initiated walk. The walks are modeled as continuous Brownian motions on the 2D plane. All strategies are measured in terms of the following metric: the decay of the probability that the query is unsuccessful with respect to t, the time duration of the query. Intuitively, the faster this decay rate, the more efficient the query, as a small time duration will suffice to locate the desired information with high probability. It is shown that the simple source-driven search decays as $(\log(t))^{-1}$; with distributed replication, it decays as approximately t^{-1}; while, with the "sticky" search, the decay is given as $t^{\frac{-5k}{8}}$. Thus the sticky search even outperforms distributed replication, so long as the number of push/pull strands is at least 2. Thus, this study provides analytical support for the rumor routing and comb–needle approaches discussed above.

9.5 Data-centric storage and retrieval

Another approach to data-centric networking is to decouple the sensor data storage location from the location where the data are generated. In this structured approach, storage location is carefully chosen based on the type and value of the corresponding data. Instead of blind querying, this enables more efficient retrieval of desired information. It also avoids the overheads associated with pure push-based schemes, where all the data are sent to the sink, regardless of whether they are needed.

9.5.1 Geographic hash tables (GHT)

The use of geographic hash tables [172] provides a simple way to combine data-centric storage with geographic routing. It is quite simple in essence and works as follows. Every unique event or data attribute name that can be queried for is assigned a unique key k, and each data value v is stored jointly with the name of the data as a key value pair (k, v). Two high-level operations provided are Put(k,v), and Get(k). A geographic hash function is used to hash each key to a unique geographic location (x, y coordinate) within the sensor network coverage region. The node in the network whose location is closest to this hashed location (known as the home location for the key), is the intended storage point for the data. When a sensor node generates a new value, the Put operation is invoked, which uses the hash function to determine the corresponding unique location and uses the GPSR geographic routing protocol to route the information to the home node. When the sink(s) issue a Get(k) query, it is sent directly to the same location.

   To ensure that the geographic routing consistently finds the same node for a key, and to provide robustness to topology changes, a perimeter refresh protocol is provided in GHT. To provide load balancing in large-scale networks, particularly for high-rate events, GHT also provides a structured replication mode. In this mode, instead of a single location, for each unique key a number of symmetric hierarchical mirror locations are chosen throughout the network. When a node generates data corresponding to the key, it stores it at the closest mirror location, while queries are propagated to all mirror locations in a hierarchical manner

9.5.2 Distributed index for multi-dimensional data (DIM)

DIM [119] is a storage and retrieval mechanism uniquely geared towards multi-dimensional range queries. An example of a multi-dimensional query is "list events such that the temperature value is between 20 and 30 degrees, and light reading is between 100 and 120 units." It comprises two key mappings:

1. All multi-dimensional values are mapped (many-to-one) to a k-bit binary vector.
2. Each of the 2^k possible binary codes is mapped to a unique zone in the network area.

Assume that all values are normalized to be between 0 and 1. The k-bit vector is generated by a simple round-robin technique. If the data are m-dimensional, the first m bits indicate whether the corresponding values are below or above 0.5, the second m bits whether the corresponding values are in the ranges [0–0.25, 0.5–0.75] or in the ranges [0.25–0.5, 0.75–1] (with disambiguation within the ranges provided by the first set of bits), and so on. Consider two examples: let $k = 4$, $m = 2$, the value (0.23, 0.15) is denoted by the binary vector 0000 (which fits all values in the multi-dimensional range (0–0.25, 0–0.25); and the value (0.35, 0.6) is denoted by 0110 (which fits all values in the multi-dimensional range (0.25–0.5, 0.5–0.75).

The mapping of binary codes to zones in a rectangular 2D network area A is performed by the following simple splitting construction: for each successive division, split the region A into two equal-size rectangles, alternating between vertical and horizontal splits. Each division corresponds to a successive bit. If the split-line is vertical, by convention, a "0" codes for the left half, and if the split-line is horizontal, a "0" codes for the top half. This construction, illustrated in Figure 9.8, uniquely identifies a zone with each possible binary vector. In a manner similar to GHT, the node closest to the centroid of the corresponding zone may be regarded as the home node, and treated as the unique point for storage and retrieval.

9.5.3 Distributed index for features (DIFS)

DIFS [67] is a technique suitable for index-based storage and retrieval of information in response to range queries (e.g. "did any sensors report readings of

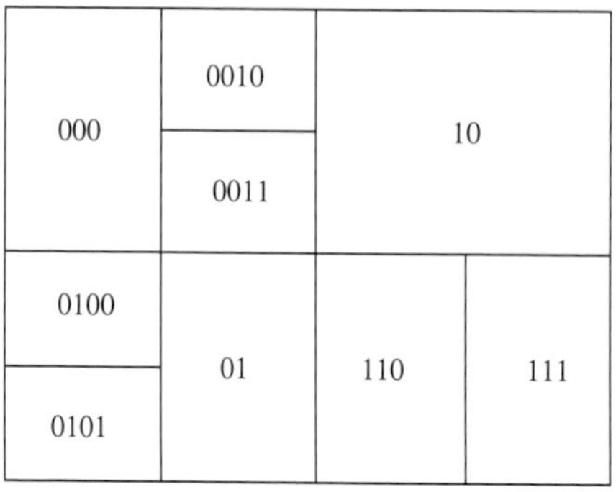

Figure 9.8 Zone creation based on binary index for storing multi-dimensional range data

temperatures within 20–30 degrees?"). DIFS constructs a multiply rooted hierarchical index as follows. Nodes store information for a range of values in a given geographic region. Nodes at low levels cover a wide range of values within a small region, while nodes at the higher levels cover a small range of values within a larger region. In DIFS, each parent has exactly four children, while each child has k parents. Each parent holds information on $1/k$ of the values that a child does, but covers four times its geographic range. A source node measuring an event sends it first to the nearby local index node (determined by a suitable hash function) with a small area coverage and largest range of values. This node then propagates a histogram of observed values to the particular parent at the next higher level with a smaller range of values covering that value, and so on. The leaf index nodes at level 0 point directly to storage nodes, while nodes at level 1 and higher each store four histograms pointing to each of the lower-level index nodes covering smaller areas. DIFS searches may enter at any level of the index structure (and often at multiple points), depending on the spatial extent and value range requested in the query, and drill down to obtain events satisfying the query. The histograms are also useful in resolving more sophisticated queries involving distributions.

9.5.4 DIMENSIONS

A multi-resolution storage and retrieval functionality suitable for spatio-temporally correlated data is provided by the DIMENSIONS architecture [59, 60]. DIMENSIONS incorporates three key components:

1. Multi-resolution hierarchical storage: In DIMENSIONS, the lowest levels of the hierarchy store high-resolution information, while the highest levels store lossy compressed coarse-grained information. Specifically, at the lowest

level of the hierarchy, individual nodes store time series data, possibly with local lossless compression. At each progressively higher level, nodes receive lossy compressed data from multiple children that they uncompress, combine together, and compress to a higher lossy compression ratio, using wavelet compression to send up to the node at the next level. Thus the nodes at higher levels have information about the larger geographic range, but at a coarse grain, while nodes at lower levels have information about smaller regions at a finer granularity.

2. Drill-down querying: Over this hierarchical structure, queries are resolved in a drill-down manner from the top. First responses at the coarsest grain are used to determine which low-level branch is likely to resolve the query, and this process is repeated until the query moves sufficiently down the structure to be resolved.

3. Progressive aging: In practice, such a storage system will face the practical limitation of finite storage. The design principle advocated for such storage in DIMENSIONS is the concept of *graceful aging*. The more extensive fine-grained data at the lowest levels of the earlier hierarchy age and are replaced with incoming new data faster, while the coarse-grained compressed information at higher layers is replaced more slowly. Thus, the farther back in time the data being queried for, the more likely it is that they will be obtained in a summarized form; queries for more recent data are answered with finer granularity.

9.6 The database perspective on sensor networks

So far we have been focusing on a bottom–up networking-centric view of WSN, with an emphasis on networking mechanisms such as routing and query forwarding, albeit enhanced with data-centric notions of in-network processing, data management, storage, and efficient retrieval of queried information. A complementary, top–down perspective is to view sensor networks as a distributed database system, and place the primary emphasis on the interface between the end user (i.e., the human querying the system) and this system. Several researchers have advocated this alternate perspective [13, 129, 233, 66].

9.6.1 Query language

It has been shown that a simple SQL-like declarative language with some extensions can be quite powerful for phrasing a diverse set of queries relevant

to sensor networks. The canonical example of such a query approach is the
TinySQL/TinyDB implementation of a tiny aggregation service (TAG) [129].

In TinyDB/TAG, all nodes in the network are treated as forming a single
table, called sensors. The columns of the table are all the attributes defined
for the network and application (e.g. these may be metadata like nodeID, loca-
tionID, timeStamp, as well as data readings from the different sensors, such as
temperature, light, etc.).

Basic queries in TinySQL can be phrased in the following manner:

```
SELECT {aggregates, attributes} FROM sensors
  [WHERE {selPreds}]
  [GROUP BY {attributes}]
  [HAVING {havingPreds}]
  [EPOCH DURATION {duration}]
```

A simple example is:

```
SELECT max{temperature}, locationID FROM sensors
  WHERE lightIntensity > 120
  EPOCH DURATION 30s
```

The query thus formulated is then flooded through the network and the pertinent information is sent from the nodes that satisfy the relevant predicates.
Aggregation operations are applied within the network en route, as the data return
to the querying sink (more on this below).

The basic TinySQL language is already quite expressive, enabling users to
formulate a wide variety of useful queries. Further enhancements have been
proposed [78]. The first is the addition of an event-driven query mechanism,
allowing a construct like:

```
ON EVENT fire-detected
SELECT max{temperature}, locationID FROM sensors
  ...
```

The query is then activated only when the named event occurs. Another
enhancement is to allow for storage of buffered data locally within the network.
It is demonstrated that, with these slight changes, the expressive power of the
language is increased greatly. For example, an illustration given in [78] is a
simple high-level program consisting of less than 25 lines that tasks the whole
network to track a moving target with query handoffs (i.e. only nodes near the

target are activated and provide information on it as it moves). This suggests the possibility of easy high-level programming of sensor networks using an SQL-like language.

9.6.2 Aggregate queries

In TAG, the responses to queries are routed up a tree, with aggregation operators such as MAX, SUM, COUNT etc. applied at each step within the network. Aggregates are implemented via three functions: a merging function f, an initializer i and an evaluator e; e.g., if f is AVERAGE, then given partial state records $PSR1 = < S_1, C_1 >$ and $PSR2 = < S_2, C_2 >$ (S and C for sum and count respectively), $PSR3 = f(PSR1, PSR2) = < S_1 + S_2, C_1 + C_2 >$. The initializer i gives the partial state record for a single sensor reading; e.g., if the single sensor value is x, then $i(x)$ returns $< x, 1 >$. The evaluator e performs the computation on a state record to return the final result; e.g. $e(< S, C >) = S/C$.

The communication savings due to aggregation within the network depend very much on the type of aggregate used. Aggregates such as COUNT, MIN, MAX, SUM, AVERAGE, MEDIAN, HISTOGRAM, etc. all have different behaviors. A classification of these aggregates along multiple dimensions, such as duplicate sensitivity and the size of partial state records, is given in [129] and used to compare aggregation performance.

9.6.3 Other work

There are several other interesting works pertinent to the database perspective on WSN. In GADT [50], a probabilistic abstract data type suitable for describing and aggregating uncertain sensor information is defined. The temporal coherency aware network aggregation (TiNA) technique [194] provides for additional communication optimization by providing temporal aggregation – data values that do not change from the previous value by more than a tolerance level θ are not communicated. Shrivastava *et al.* propose and analyze data aggregation schemes suitable for medians and other more sophisticated aggregates, such as histogram and range queries [197]. The problem of aggregation operators over lossy networks is addressed by Nath *et al.* [144], who provide an analysis of synopsis diffusion techniques that provide robust order and duplicate insensitive aggregation that is decoupled from the underlying multi-path routing structure.

Yao and Gehrke [233] discuss taking into account available metadata about the state of different parts of the network to provide an optimized query plan distributed across query proxies on sensor nodes. The query plan describes both the data flow within the network as well as the computational flow within the node. Bonfils and Bonnet [12] address the problem of optimizing the placement of query operators for long-standing queries autonomously within the network through an exploratory search process.

9.7 Summary

Unlike traditional communication networks that must support a wide range of applications (some not even known at design time), WSNs are much more application specific in nature. Communication in a WSN is most often pertinent to the information available at sensors or desired by an external user. A data-centric approach, where the routing is based on named data rather than addresses, can be advantageous for two reasons: (a) it eliminates the overhead associated with name binding and (b) it allows for energy efficiency through in-network processing, including compression and aggregation of information. The directed diffusion routing mechanism is unique in routing based on named attributes rather than traditional IP-style addressing.

Several studies, including the cluster-based LEACH protocol and many analytical studies, have examined the problem of routing with in-network compression in sensor networks. The studies suggest that, while finding optimal joint routing–compression routes may be difficult, good approximations are possible. It is possible to achieve near-optimal energy performance for routing with compression with a simple LEACH-like clustering technique that is not correlation aware.

Besides end-to-end routing, data discovery and querying form an important communication primitive in sensor networks. Alternatives to the high-overhead naive flooding approach are desired. Several querying techniques have been proposed and analyzed, including expanding ring search, IDSQ, and ACQUIRE. Rumor routing and the comb–needle approach both advocate hybrid push–pull rendezvous techniques, where query trajectories from sinks intersect with event notification trajectories from sources, and show that they can offer significant gains.

Data-centric storage techniques including GHT, DIM, and DIFS offer another alternative by decoupling the location of data storage from the location where data are generated. Data are indexed at locations that depend upon the named content, which makes retrieval much easier with lower overheads than blind querying. The DIMENSIONS technique advocates multi-resolution storage with graceful aging so that more recent fresh information is available at a finer granularity than older data.

Finally, a complementary perspective to data-centric routing and storage in sensor networks is to treat them as distributed databases. TinyDB is a key effort in this direction, advocating the use of a simple-yet-powerful SQL-based declarative query language and in-network processing of aggregate queries.

Exercises

9.1 *Analysis of push versus pull:* Consider a very simple mathematical model to analyze push diffusion and pull diffusion routing. Time is divided into multiple epochs. In each epoch, with probability p_D there is one active source in the network (unknown to the sinks) that generates data, and with probability p_I there is one active sink in the network (unknown to the sources) that is interested in the data. In pull mode, the active sink floods its interest to all n nodes, and, if there is an active source, it responds directly with data along the shortest path to the active sink. In push mode, the active source floods an event notification to all n nodes, and, if there is an active sink, it responds with a reinforcement message down the shortest path to the active source, which can then send data along this path to the active sink. Assume the shortest path between an active source and an active sink is always of length $\sqrt{n}$. Assume interest messages are of size I, event notification messages are of size N, and data messages of size D. Derive an expression for the condition under which push incurs less overhead than pull, and explain it intuitively.

9.2 *Expanding ring search:* Consider an $n \times n$ grid of sensor nodes, where each node communicates with only its four neighbors. Queries are issued for a piece of information located at a random node in the network by a node at the center through a hop-by-hop expanding ring search.

(a) Derive an expression for the expected number of steps until query resolution.

(b) What is the expected number of steps when the information is located at two random nodes (instead of one)? Comment.

9.3 *Trajectory-based forwarding:* How should a circle be represented in parametric $(x(t), y(t))$ form? Simulate the deployment of a 100 node random $G(n, R)$ network with $R = 0.2$ in a unit area. Using any convenient forwarding rule, show the nodes visited by a TBF query that aims to follow the big in-circle centered in the middle of the area with radius 0.5.

9.4 *Rumor routing:* Simulate rumor routing on an arbitrary network of n nodes using random walks as trajectories. Vary the number of source- and sink-initiated walks and quantify the tradeoff between energy cost and latency of query resolution with increasing numbers of walks.

9.5 *DIM binary code mapping:* Give the binary codes that correspond to the following values if $k = 8$: (a) (0.23, 0.15), (b) (0.35, 0.6), (c) (0.83, 0.29).

9.6 *DIM zone creation:* In a square area, draw the regions that correspond to the following codes: (a) 1010, (b) 1101, (c) 00001.

Transport reliability and congestion control

10.1 Overview

The objective of transport layer protocols is to provide reliability and other quality of service (QoS) services and guarantees for information transfer over inherently unreliable and resource-constrained networks. The following are the interrelated guarantees and services that may be needed in wireless sensor networks:

1. **Reliable delivery guarantee:** For some critical data, it may be necessary to ensure that the data arrive from origin to destination without loss.
2. **Priority delivery:** The data generated within the WSN may be of different priorities; e.g., the data corresponding to an unusual event detection may have much higher priority than periodic background readings. If the network is congested, it is important to ensure that at least the high-priority data get through, even if the low-priority data have to be dropped or suppressed.
3. **Delay guarantee:** In critical applications, particularly those where the sensor data are used to initiate some form of actuation or response, the data packets generated by sensor sources may have strict requirements for delivery to the destination within a specified time.
4. **Energy-efficient delivery:** Energy wastage during times of network congestion must be minimized, for instance by forcing any necessary packet drops to occur as close to the source as possible.
5. **Fairness:** Different notions of fairness may be relevant, depending on the application. These range from ensuring that all nodes in the network provide

equal amounts of data (e.g. in a simple data-gathering application), to max–min fairness, to proportional fairness.

6. **Application-specific, data-centric quality of service:** In general, a data-centric QoS goal requires that the network as a whole provide as accurate a picture of the sensed environment as possible, given the bandwidth/energy resource constraints. In some applications, it may suffice just to ensure that some number of reports about specified events arrive at the destination in each unit time, regardless of which exact sources send the information.

In an ideal system, with large bandwidths and very low loss rates, the reliability and QoS guarantees being sought would not be difficult to provide. The task of designing transport mechanisms to provide the above functionalities is challenging because of a number of practical factors and limitations:

1. **Channel loss:** Due to signal decay and multi-path fading effects, as we have seen before, the error rates may be quite significant in WSN links. Further, these loss rates are very much a function of the exact location of the nodes as well as the environment, and can therefore fluctuate greatly over space and time.

2. **Interference:** The error rates on wireless links are also very much a function of the level of interfering traffic in the vicinity. When traffic rates are high, packet losses can occur, even on otherwise good-quality links when the SINR drops below the successful reception threshold because of interference.

3. **Bandwidth limitation:** Even if individual sensors generate data at a low rate compared with the maximum data rate available, the aggregate traffic of a large-scale network can be large enough to cause congestion in these low bandwidth networks. This happens particularly when all traffic is headed in the same destination, resulting in bottlenecks near the sink.

4. **Traffic peaks:** While the average data rate in a WSN is low most of the time, there may be a much higher peak traffic rate whenever important events are being detected. Such traffic peaks are likely to be highly correlated in both space and time – resulting in congested *hot-spots*.

5. **Node resource constraints:** The intermediate nodes in the network may also suffer from other constraints that can impact transport techniques: (i) low computational processing capabilities that preclude high-complexity approaches, (ii) memory/storage constraints that limit the size of the message buffer (causing packet losses during congestion events), and, as always, (iii) energy constraints that limit the amount of possible transmissions.

There are many ways in which transport layer challenges in a sensor network differ from those for the traditional wired Internet:

- The primary traffic pattern of a many-to-one nature, as opposed to one-to-one, flows between arbitrary hosts.
- Wireless links (a) have time-varying, potentially very high, error rates due to fading effects, and (b) have to share the available bandwidth with other nearby links.
- Energy efficiency is a serious concern – lost packets not only reduce through-put, but (depending on where they occur) also cause energy wastage.
- The quality-of-service requirements may not be specified on an individual source-destination address-centric basis, but in a data-centric manner (e.g., by requiring that sufficient quality of information about an event be provided to the sink, regardless of which nodes originate that information).

For these reasons, we shall see that many of the proposed transport mechanisms are fundamentally different from the traditional TCP-like approach of implementing congestion control at the end hosts alone. Indeed most of the transport mechanisms proposed for WSN are implemented in a distributed manner hop-by-hop at each step of the way, in intermediate nodes of the network.

10.2 Basic mechanisms and tunable parameters

Given the functionalities that need to be provided and the constraints that make this a challenging task, we must examine the various parameters that can be tuned by a WSN transport protocol.

1. **Rate control:** This refers to changing the rate at which packets are sent by a node, for instance by introducing a tunable wait time before which a packet is sent to the link layer for transmission. The rate control may be done purely at the source nodes, or at the intermediate nodes. In the latter case, there may be different rates for the route-through traffic and the traffic originating at the given node. For implementations requiring fairness, the route-through traffic may have to be further divided on a per-child basis.
2. **Scheduling policy:** The transport buffer may be a regular first-in first-out queue (FIFO) in the simplest case. But, to support fairness, multiple FIFO queues served in a round robin fashion may be used. Similarly, to support priority and delivery guarantees, priority queues may be needed. These may be also be approximated by multiple prioritized queues to minimize complexity.

3. **Drop policy:** The queue drop policies can also have a significant impact on transport guarantees. The most basic technique is the traditional drop-tail policy employed commonly with FIFO queues whereby the packets are dropped from the tail of the queue. However, a more sophisticated policy may choose to drop low-priority packets, or even high-priority packets that cannot reach the destination before a required deadline. The drop policy may also take into account the distance travelled by a packet – it may be preferable to drop a packet near its origin as opposed to a packet that has travelled a great distance, for reasons of energy efficiency and fairness.

4. **Explicit notification:** Intermediate nodes may monitor their queue lengths and when a threshold is exceeded send explicit signals to their children nodes to reduce the rates. This back pressure is typically propagated down to the sources to reduce rates. Similarly, if the queue length is sufficiently small, signals may be sent to increase rates. In other schemes, the sink itself may generate explicit notification asking nodes within the network to increase or decrease their traffic, depending upon whether application QoS needs are satisfied.

5. **Acknowledgements:** One approach to ensuring reliable transport is to use a sequence of packets and send negative acknowledgements (NACKs) hop-by-hop to trigger retransmissions for loss recovery. In some cases it may be better to use link-layer acknowledgements alone, but exercise some control over the maximum number of retransmission attempts, depending upon the priority of the packet.

6. **MAC backoff:** Some of the transport mechanisms directly exert influence over link layer parameters, such as the backoff level. For instance, higher-priority packets may be given a shorter backoff interval. Another approach is to introduce an additional random backoff as a jitter to ease congestion in hot spots.

7. **Next-hop selection:** Technically, forwarding decisions should be left to the network layer alone. However, sometimes the satisfaction of strict transport-level QoS objectives (such as in the case of realtime applications with strict deadlines) may require an integrated cross-layer approach, whereby the congestion state information is used to determine the next hop.

10.3 Reliability guarantees

We begin by examining various techniques that have been proposed to ensure reliable delivery of information in WSN.

10.3.1 Pump slowly fetch quickly (PSFQ)

In some contexts such as over-the-air reprogramming/retasking of all nodes in a WSN it is essential to ensure that all the data being transmitted are received perfectly error free at the intended destinations. The *Pump Slowly Fetch Quickly* (PSFQ) paradigm [215] is proposed for this scenario. Note that contrary to the typical source to sink flows for data-gathering, in this context the traffic flows out from the sink to the multiple network nodes. It is assumed that the code is being propagated after fragmentation into smaller component segments.

It is essential that no segment be lost to any destination, as the code would be useless if even a small piece of it were missing. PSFQ aims to provide lossless delivery, minimize the overheads of control messages, and also provide some form of delay guarantees for data delivery to all intended receivers. The key idea of PSFQ is that the propagation of each code segment is done at a slow rate. In between the propagation (pump) of each segment, there is sufficient time for intermediate nodes to detect, based on sequence numbers, that a loss has occurred, and, if so, to request hop-by-hop retransmissions within the network (fetch). This is illustrated in Figure 10.1.

The authors of PSFQ [215] argue that because link error rates can be high, and error propagation would be exponential over multiple hops, an end-to-end approach to error recovery is likely to incur a significantly greater overhead than hop-by-hop retransmissions. The authors also present simple simulations suggesting that the benefits of allowing retransmissions show diminishing returns after five retries and use this as the ratio of pump to fetch operations.

When a node correctly receives a segment with a sequence number that is more than one sequence number higher than the previously received segment, a loss

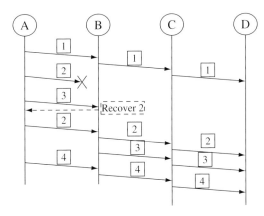

Figure 10.1 Data recovery in PSFQ

event is detected. When a loss event is detected, the node does not forward any segment farther until the missing segment is fetched, to localize the loss event and minimize the control overhead from downstream nodes. Loss aggregation is employed to fetch multiple lost segments together. PSFQ also enables multiple neighbors to send a missing segment to a requesting node (with contentions resolved by the use of a random timer).

10.3.2 The reliable multi-segment transport mechanism (RMST)

Stann and Heidemann [206] compare several reliability mechanisms in developing RMST, including the following: (i) ACK-based ARQ for unicast packets at each hop, (ii) end-to-end NACK-based recovery, and (iii) hop-by-hop NACK-based recovery. The first is a purely link layer approach; the second is a pure end-to-end mechanism: only the sink detects loss based on sequence numbers and requests the appropriate source to retransmit the message. The third is similar to PSFQ: each node along the route caches successful packets, checks for missing packets, and requests repair by retransmission from the local cache of the node before it.

Based on extensive simulations of these approaches, the authors of RMST find, as with PSFQ, that hop-by-hop recovery is more efficient than end-to-end recovery. However, they also suggest that the hop-by-hop NACK-based recovery should be built on top of, or in addition to, a purely link-layer ACK-based ARQ.

10.4 Congestion control

Because of the low available bandwidth in WSN, congestion events are quite likely, particularly during peak traffic due to event detections. Several researchers have proposed different solutions for WSN congestion control.

10.4.1 Adaptive rate control (ARC)

The adaptive rate control technique [224] aims to provide a simple adaptive distributed rate control to ensure fairness and provide congestion control for a data-gathering tree. First, a distinction is drawn between originating traffic (i.e. traffic originating at the given node) and route-through traffic (i.e. traffic passing through that node that originated in upstream nodes below it on the tree).[1]

---

[1] The terminology of upstream/downstream can sometimes be confusing. Consider the one-way data-gathering, where the sink is the root to which all flows are directed. A node i is the parent of node

If the application rate of originating data is S, then the output originating rate is $S \cdot p$ where $p \in [0, 1]$. Depending on the conditions, the rate is increased linearly by an amount α and reduced multiplicatively by a factor $\beta \in (0, 1)$. Route-through traffic is given preference via a less aggressive reduction $\beta_{\text{route}} = 1.5\beta_{\text{originate}}$, and fairness is provided by ensuring that $\alpha_{\text{originate}} = \alpha_{\text{route}}/(n + 1)$, where n is the number of descendant nodes generating the route-through traffic. Implicit acknowledgements are obtained through listening when the node on the next hop retransmits an earlier forwarded message – this signal is used to control the increase/decrease functions. Thus, while no explicit signals are sent, a congestion event would cascade down through a series of originating and route-through rate reductions, acting as an implicit back-pressure that slows the rates at all points below the location of congestion.

Another contribution of this work is to suggest the introduction of a jitter through a random initial backoff at the MAC level, to minimize contention/collisions of synchronized traffic generated by sensors sensing a common event.

10.4.2 Event-to-sink reliable tranport (ESRT)

The event to sink reliable transport mechanism [181] takes a data-centric view of the problem of congestion control. It is assumed there are nodes in a given event region sending messages frequently. From an application point of view, it is assumed that there exists some threshold total reporting rate (number of reports received in a decision interval from any of the sources) that is needed for reliable event detection.

ESRT is a closed-loop rate control technique. It is assumed that the sink node has a high-power radio that can be used to provide feedback to the sources directly. When the number of received packets is plotted as a function of the reporting frequency as in Figure 10.2, it shows a maximum at some point $f_{\max}$, then declines due to congestion. When the sink observes that the reliability is below threshold and the reporting rate is below $f_{\max}$, it signals the nodes to increase their rate to meet the reliability criterion. If the reliability is above threshold, the reporting rate is reduced to save energy. In case the network is congested and the reliability is below threshold, the reporting rate is reduced aggressively. If the network provides a reliability that is within ϵ (which could be about 1%) of the reliability threshold, and no congestion is taking place, then no change in

---

j if it is the next hop from j towards the sink. We then refer to node j as being *below* i on the tree; however, we also say that node j is *upstream* from i, because its data are flowing towards the sink through i. Similarly node i is *above* j, but it is *downstream* from j.

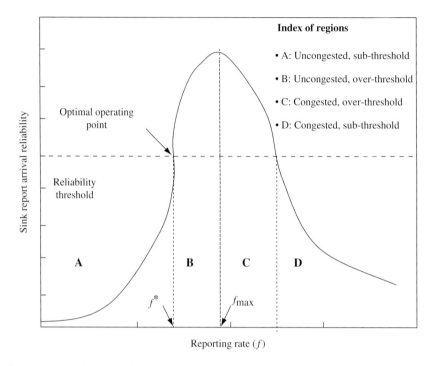

Figure 10.2 Sink report reliability curve used for congestion control in ESRT

reporting rate is required. The signals from the sink are transmitted once every decision period. The sink is informed that congestion is taking place by the nodes in the network, which detect it by measuring local buffer occupancy and flag a congestion bit in packets sent towards the sink.

10.4.3 Congestion detection and avoidance (CODA)

CODA [216] is a congestion control technique for WSN that comprises three mechanisms:

1. Congestion detection: Buffer occupancy does not give a good indication of congestion if ARQ is not used, because the queue can potentially clear even if packets are being lost due to collision. It is also possible for nodes to determine congestion by listening to the channel and to determine how busy/loaded it is; however, this can have a significant energy cost. CODA therefore uses a periodic sampling technique with exponential averaging, to mitigate the impact of temporary fluctuations.
2. Open-loop hop-by-hop back-pressure: When congestion is detected, a suppression message (an explicit congestion notification) is propagated upstream

towards the source. The suppression message is sent repeatedly. Nodes can respond to this message by dropping packets or reducing their rate. The suppression message can also be used as the basis of an implicit priority scheme, whereby only data of lower priorities are suppressed. Care must be taken to ensure that the suppression message propagates all the way upstream, or else nodes upstream will not be aware of the congestion.

3. Closed-loop multi-source regulation: When the congestion is persistent, it is helpful for the sink to play a role, by providing feedback for rate control. This is similar in spirit to ESRT. When the source event rate is higher than some fraction of the maximum throughput, the source may contribute to increased congestion by transmitting congestion notifications repeatedly. Therefore, in this case, the source flags a "regulate" bit to indicate to the sink that it is in this high-congestion regime. The sink then responds with periodic ACKs covering multiple event reports to regulate all sources associated with that event. If it finds the report rate to be lower than expected, it stops sending ACKs until the congestion clears, so that the sources can reduce their rate. In general, sources maintain, decrease, or increase their rate depending on how frequently they receive these ACKs.

10.4.4 Fair rate allocation

An explicit approach to congestion control with ensured fairness is given by Ee and Bajcsy [42]. Their mechanism comprises the following three steps:

1. Determine the average transmission throughput r. One mechanism to do this is to measure the rate as the inverse of the time interval to transmit a single message. The interval is measured from the time when the transport layer sends the packet to the network layer to the time when the network layer reports that the packet has been transmitted. The rate measurement uses an exponential moving average.

2. Divide r among the number n of upstream devices (i.e. on the subtree below the node on the data-gathering tree), so that the nominal per-node data packet generation rate is $r_{\text{data}} = r/n$. The size of the subtree is easily determined through a simple count technique. If the queue is full or overflowing, however, the rate is reduced to an even lower value.

3. Obtain the rate of the node's parent $r_{\text{data,parent}}$ through promiscuous listening or via a control message. Compare r_{data} with $r_{\text{data,parent}}$ and propagate the smaller rate upstream to the nodes in the subtree.

Fairness is obtained by measuring and dividing the rate by the number of downstream nodes. To implement this, every node maintains separate FIFO queues for each child node as shown in Figure 10.3. Then, a probabilistic selection mechanism is employed to weigh the choice of packets so that the probability of choosing a queue from which to transmit a packet is proportional to the number of nodes serviced by that queue.

10.4.5 Fusion

The fusion technique [93] is actually a combination of three techniques to mitigate congestion:

1. Hop-by-hop flow control: this technique is very similar to the open-loop flow control in CODA.
2. Source rate limiting: this is similar in spirit to the ARC technique, implemented slightly differently, as follows. It is assumed that all nodes offer the same traffic load (an assumption that can be relaxed with the addition of more information collected about different node rates). Each sensor listens to the traffic its parent forwards to estimate N, the total number of unique sources routing through that parent. A token bucket scheme is then employed whereby a node gathers a token once it hears its parent forward N messages. The sensor can transmit its own generated data only if its bucket (which has a maximum token limit) contains at least one token. This approach ensures that the sensor sends its own data fairly, at the same rate as each of its descendants.
3. Congestion-adaptive backoff control: The length of each node's randomized backoff is determined as a function of its local congestion state. A sensor node that is congested uses a shorter backoff window, allowing it to win the contention period with high likelihood. This will help backlogged queues drain faster.

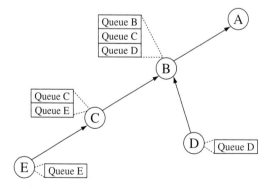

Figure 10.3 Multiple FIFO queues for fair delivery

Fusion is evaluated on a testbed of 55 motes. Through an extensive experimental evaluation, the fundamental need for congestion control mechanisms is demonstrated (as otherwise efficiency and fairness are severely degraded). It is found that as loads increase or when channel variations cause reduction in bandwidth, nodes must reduce send rates so that the network avoids congestion collapse. The fusion study also shows that the hop-by-hop flow control and the backoff control with simple queue occupancy-based congestion detection, work well together under a wide range of workloads, and the rate control provides a substantial degree of fairness.

10.5 Real-time scheduling

In some critical applications, particularly those involving control and actuation elements as well as emergency response activities, data communications in a WSN may include strict time constraints in the form of end-to-end deadlines, e.g. the information detection of a security breach in a perimeter surveillance application may need to be conveyed to the sink within 10 seconds so that appropriate measures can be taken. In this case, the reliability/QoS objective of the network shifts from maximizing the delivery ratio (the fraction of packets delivered successfully) to minimizing the packet deadline miss ratio (the fraction of packets delivered that missed their deadlines).

10.5.1 Velocity monotonic scheduling (VMS)

RAP is an architecture for realtime communications [125] that includes velocity monotonic scheduling (VMS), an interesting velocity-based approach to address this issue. VMS is a packet scheduling policy that determines the order in which incoming packets are forwarded. VMS takes into account the time t till the deadline of the packet expires as well as the physical distance d between the current node and the destination node. The required velocity for this packet to reach the destination is then d/t. The velocity can be calculated statically at the source, or dynamically at each intermediate hop as it progresses through the network. Each packet is then assigned a priority that is monotonic with respect to its required velocity (i.e. packets with a higher required velocity have a higher priority). This is illustrated in Figure 10.4. In the general case, the decision depends only on the estimated velocity. One special case is when packets are intended for the same destination; in this case they will be scheduled using an earliest-deadline-first discipline. Another special case arises when packets are

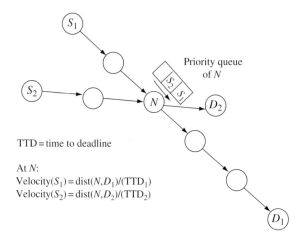

TTD = time to deadline

At N:
Velocity(S_1) = dist(N,D_1)/(TTD$_1$)
Velocity(S_2) = dist(N,D_2)/(TTD$_2$)

Figure 10.4 Velocity monotonic scheduling

intended for different destinations with the same deadline; in this case, they will be scheduled using a farthest-destination-first discipline.

VMS can be implemented using priority queues. However, since priority queues can have an unbounded insertion time (growing as a logarithmic function of the queue size), this can also be approximated by having separate FIFO queues for different priorities. The authors also propose that queues drop packets that have already missed their deadlines to avoid wasting bandwidth.

Finally, RAP/VMS also coordinates the priority for packet transmissions across multiple nodes by controlling the random backoff mechanism at the MAC layer. Thus packets with higher priority use a smaller backoff, allowing them to wait shorter periods of time for medium-access.

10.5.2 Velocity-based forwarding (SPEED)

Another approach to real-time delivery is to use a velocity calculation to determine the forwarding choice. The SPEED protocol [73] implements such a mechanism. In SPEED, each node i estimates the delay for each of its neighbors $t_{i,j}$ (the time it would take them to transmit a packet that is forwarded to them) through a periodic RTT calculation with exponential averaging. Let L_i be the distance of any node i from the destination. Then the node calculates the relay speed $v_{i,j}$ of a neighbor j as

$$v_{i,j} = \frac{L_i - L_j}{t_{i,j}} \tag{10.1}$$

The essence of the SPEED algorithm is that, when it has a packet to route, it picks a forwarding node from the set of neighbors that are closer to the destination than itself for which the calculated speed is higher than some threshold v_t. This selection could be deterministic, always picking the neighbor with highest relay speed, or probabilistic, to provide some load balancing, e.g. with an exponential distribution depending on the candidate neighbors' relay speeds. In this way, SPEED aims to provide an approximate delay guarantee of L_{max}/v_t, where L_{max} is the maximum distance from any node in the network to the destination. This process provides an implicit way of providing rerouting during periods of congestion.

However, due to congestion, it can happen that there is no neighbor of the node that has a relay speed higher than v_t and is closer to the destination than itself. Therefore SPEED also includes a neighborhood feedback loop (NFL) that is crucial to maintaining relay speeds. It then looks at all neighbors closer than itself to the destination and forwards to them probabilistically. The ratio of the number of packets that are lost or sent to a neighbor with relay speed less than v_t to the total number of packets sent is called the miss ratio for that node m_i. The NFL consists of a local relay ratio controller that takes the neighboring nodes' miss ratios to determine what fraction of packets to relay and what fraction to drop, called the relay ratio r_i. Formally, let n_i be the number of neighbors of node i, $\sum m_j$ the sum of the miss ratios of its neighbors, and K a proportional gain; then the relay ratio r_i is given as

$$r_i = 1 - K \frac{\sum m_j}{n_i} \tag{10.2}$$

The effect of this policy is to reduce the rate of a node when the downstream nodes are congested. To reduce the rate of nodes further upstream from it, an explicit suppression back-pressure signal is also generated.

10.6 Summary

Transport-layer mechanisms are needed in wireless sensor networks to provide guarantees on reliable, low-latency, energy-efficient, fair delivery of information. Several challenges must be overcome in order to provide these guarantees: channel loss, interference, bandwidth limitations, bursty traffic, and node resource constraints.

For reliable data delivery, unlike the traditional Internet, end-to-end TCP-based mechanisms are not appropriate, due to the high loss rate on links. Hop-by-hop sequence-based loss detection and NACK-based ARQ recovery mechanisms are advocated by both the PSFQ and RMST studies.

Because of limited and fluctuating bandwidth resources, in large-scale networks, even if individual per-source data rates are low, congestion is likely to occur. Several related studies and protocols have focused on congestion control in WSN, including ARC, ESRT, CODA, fusion. These studies suggest a number of design principles: (a) congestion may need to be estimated through a combination of queue monitoring and channel sampling, (b) back-pressure-oriented rate control techniques that try to reduce rates as close to the source as possible are desirable for energy-efficiency reasons, (c) backoff values should be adapted based on congestion states, allowing backlogged nodes to access the medium and hence clear faster, and, finally, (d) the sink can provide useful feedback to sources for rate control based on the aggregate received rate or other application-specific QoS metrics.

In time-critical WSN applications, low-latency delivery is an important goal. The techniques for real-time delivery, RAP/VMS and SPEED, advocate tracking the status of individual packets within the network to make appropriate scheduling and forwarding decisions at each hop. Packets that miss their deadline should be dropped to reduce traffic and improve energy efficiency.

Exercises

10.1 *ARQ retransmission limit:* Derive an expression for and plot the probability that a message can be successfully delivered as a function of the raw link message-error probability for different numbers of ARQ retries (varying from 0 to 10). What do you conclude about a reasonable setting for the number of ARQ retries?

10.2 *NACKs vs. ACKs:* What are the pros and cons of using negative acknowledgements (NACKs) instead of positive acknowledgements (ACKs) for providing reliable delivery?

10.3 *Fair queues:* For the illustration in Figure 10.3, what should the probability of serving each queue be at each node, in order to ensure fair data-gathering?

10.4 *Velocity monotonic scheduling:* Consider the example of VMS in Figure 10.4. What constraint must the deadline of a packet from S_2 satisfy in order for it to have higher priority, if $dist(N, D_1) = 3$, $dist(N, D_2) = 1$, and the time to deadline for the packet from S_1, $TTD_1 = 35$?

11

Conclusions

11.1 Summary

Wireless sensor networks must be designed to meet a number of challenging requirements including extended lifetime in the face of energy constraints, robustness, scalability, and autonomous operation. The many design concepts and protocols described in the preceding chapters address these challenges in different aspects of network operation. While much additional work remains to be done to realize the potential of tomorrow's systems, even these partial solutions offer reason for optimism.

Perhaps the most important lesson to take away from the studies described in this book is that the fundamental challenges must be tackled by making appropriate design choices and optimizations across multiple layers.

Consider energy efficiency, which is perhaps the most fundamental concern due to limited battery resources. The most significant source of energy consumption in many applications is radio communication. At deployment time, energy efficiency concerns can inform the selection of an appropriate mixture of heterogeneous nodes and their placement. Localization and synchronization techniques can be performed with low communication overheads for energy efficiency. At the physical/link layers, parameters such as the choice of modulation scheme, transmit power settings, packet size, and error control techniques can provide energy savings. Medium access techniques for sensor networks use sleep modes to minimize idle radio energy consumption. Topology control techniques suitable for over-deployed networks also put redundant nodes to sleep until they are needed to provide coverage and connectivity. At the network layer, routing techniques can be designed to incorporate energy-awareness and in-network compression to minimize energy usage. Combining all these techniques, it may well be possible to make sensor network deployment with battery-operated devices last for several years.

The unreliability of wireless links, with large spatio-temporal variations in quality due to multi-path fading effects, poses another fundamental challenge. Link layer solutions include the implementation of link quality monitoring along with power control, blacklisting, and ARQ techniques. The use of link quality metrics as well as diversity-based multi-path and cooperative routing techniques also provides robustness at the network layer. Finally, rate control and prior-ity scheduling techniques enable reliable communication of critical information though the network.

Scalability concerns are addressed in protocols at all layers by emphasizing distributed and hierarchical algorithms with localized interactions. We have seen that many of these techniques are also designed to be inherently self-configuring and adaptive to changes in the environment, to meet the need of autonomous, unattended operation.

11.2 Further topics

There are additional topics in wireless sensor networks that we have not examined in detail in this book. We discuss these briefly below, and give pointers to some key papers in each topic that the interested reader may seek out for further study.

11.2.1 Asymptotic network capacity

One theoretical topic relevant to sensor networks that has received considerable attention in recent years is that of the asymptotic capacity of multi-hop wireless networks. The seminal work in this area is by Gupta and Kumar [70] who show that the per-node throughput must decrease as $\Theta(1/\sqrt{n})$ as the number of nodes n in a static network increases, because of interference limitations. Scaglione and Servetto [186] have suggested that capacity need not asymptotically tend to zero in a sensor network environment where correlations in the environment can be exploited for compression, although this appears to be contradicted by the results of Marco et al. [133]. Taking a slightly different approach, Hu and Li argue that the maximum sustainable throughput in energy-limited networks asymptotically decreases even faster than their interference-limited capacity [91].

11.2.2 Hardware, software and simulation tools

On the systems side, there have been several innovations in hardware, software, and simulation tools that have been instrumental in enabling sensor network research and development. The Berkeley mote platform [82] and the accompanying TinyOS operating system [81] have been the most popular choice for sensor network experimentation. TinyOS provides a highly efficient modular design well-suited to these devices. This operating system is implemented with the NesC embedded programming language [145], which provides a high-level, event-driven model of coding. Alternative hardware and software platforms have also been developed, including the Colorado MANTIS [1], and the Medusa MK-II from UCLA. There exist several sensor network simulation platforms, including TOSSIM [212], and EmStar [44].

11.2.3 Security protocols

Because of their pervasive and sometimes critical surveillance operation, the data collected by sensor networks must be kept private, and networks must also be protected against malicious attacks aimed at disrupting or disabling their functionality. Wood and Stankovic [227] provide a comprehensive survey of denial-of-service attacks that can affect sensor networks (ranging from tampering, jamming, and energy depletion to route misdirection and black holes), along with suggestions for combatting these attacks. Karlof and Wagner [100] analyze the vulnerabilities of several existing routing protocols and suggest suitable countermeasures. Perrig *et al.* develop a suite of cryptographic techniques optimized for the resource-constrained sensor network devices, in a framework known as SPINS [159].

11.2.4 Target tracking and collaborative signal processing

Particularly because much of the early sensor network research was funded by the US Department of Defense, surveillance applications have been studied in depth. There is a considerable literature addressing various aspects of target tracking. The IDSQ technique proposed by Chu, Haussecker, and Zhao [28] provides for energy-efficient tasking of sensors, while tracking a moving target. Pattem, Poduri, and Krishnamachari [152] quantify the energy–quality tradeoffs inherent in various target-tracking approaches. Shin, Guibas, and Zhao [196] provide a solution for identifying multiple targets. Fang, Zhao, and Guibas [48] define protocols for efficient target enumeration and aggregation. D'Costa and Sayeed [38] study the tradeoffs between decision and data fusion for target classification problems. Aslam *et al.* [4] show that a simple binary detection depending on target movement direction can yield low-error target tracking.

A related domain that has been studied is the tracking of more diffuse phenomena, such as chemical plumes. A canonical problem in this domain is detecting and tracking the edge of the diffuse phenomenon: Nowak and Mitra [150] present a hierarchical boundary estimation technique; Chintalapudi and Govindan [26] propose and compare various distributed edge detection techniques. Liu *et al.* [121] propose an effective way to track diffuse phenomena by converting lines into points in a dual-space.

11.2.5 Programming and middleware

It is important to facilitate the programming of large-scale sensor networks at a high level, and this has been a topic of growing attention. The Maté virtual machine [115] provides for efficient high-level reprogramming of TinyOS-based devices. TinyGALS [25] is an event-driven globally asynchronous and locally synchronous programming model for embedded sensor networks designed to simplify and automate code generation. Liu *et al.* [123] develop a state-centric methodology for programming collaborative signal-processing applications in sensor networks. Bakshi, Ou, and Prasanna propose the automatic synthesis of compute-intensive sensor network systems that can translate formal specifications of application functionality into a working system [6]. Several researchers have advocated visions of wireless sensor network middleware that can bridge the gap between applications and tunable network protocols [77, 175, 241].

References

[1] H. Abrach, S. Bhatti, J. Carlson, H. Dai, J. Rose, A. Sheth, B. Shucker, J. Deng, and R. Han, "MANTIS: System Support for MultimodAl NeTworks of In-situ Sensors," *Proceedings of ACM WSNA*, September 2003.

[2] Z. Abrams, A. Goel, and S. Plotkin, "Set K-Cover Algorithms for Energy-Efficient Monitoring in Wireless Sensor Networks," *Proceedings of IPSN*, April 2004.

[3] K. A. Arisha, M. A. Youssef, and M. F. Younis, "Energy-aware TDMA based MAC for Sensor Networks," *Proceedings of IEEE Workshop on Integrated Management of Power Aware Communications, Computing and Networking (IMPACCT)*, May 2002.

[4] J. Aslam, Z. Butler, F. Constantin, V. Crespi, G. Cybenko, and D. Rus, "Tracking a Moving Object with a Binary Sensor Network," *Proceedings of ACM Sensys*, November 2003.

[5] P. Bahl and V. N. Padmanabhan, "RADAR: An RF-Based In-Building User Location and Tracking System," *Proceedings of IEEE INFOCOM*, March 2000.

[6] A. Bakshi, J. Ou, and V. K. Prasanna, "Towards Automatic Synthesis of a Class of Application-Specific Sensor Networks," International Conference on Compilers, Architecture, and Synthesis for Embedded Systems (CASES), October 2002.

[7] P. Baruah, R. Urgaonkar, and B. Krishnamachari, "Learning Enforced Time Domain Routing to Mobile Sinks in Wireless Sensor Fields," *Proceedings of First IEEE Workshop on Embedded Networked Sensors (EmNetS-I)*, November 2004.

[8] M. Batalin and G. S. Sukhatme, "Coverage, Exploration and Deployment by a Mobile Robot and Communication Network," *Telecommunication Systems Journal, Special Issue on Wireless Sensor Networks*, 26, 2 (2004), 181–196.

[9] M. Bhardwaj and A. P. Chandrakasan, "Bounding the Lifetime of Sensor Networks Via Optimal Role Assignments," *Proceedings of IEEE INFOCOM*, June 2002.

[10] S. Biswas, R. Morris, "Opportunistic routing in multi-hop wireless networks," *Computer Communication Review*, 34, 1 (2004), 69–74.

[11] B. Bollobás, *Random Graphs*, Academic Press, 1985.

[12] B. Bonfils and P. Bonnet, "Adaptive and Decentralized Operator Placement for In-Network Query Processing," *Proceedings of IPSN*, April 2003.

[13] P. Bonnet, J. E. Gehrke, and P. Seshadri, "Querying the Physical World," *IEEE Personal Communications, Special Issue on Smart Spaces and Environments*, 7, 5, October 2000, 10–15.

[14] P. Bose, P. Morin, I. Stojmenovic, and J. Urrutia, "Routing with Guaranteed Delivery in Ad Hoc Wireless Networks," ACM Int. Workshop on Discrete Algorithms and Methods for Mobile Computing and Communications DIAL M, August 1999.

[15] D. Braginsky and D. Estrin, "Rumor Routing Algorithm for Sensor Networks," *Proceedings of ACM WSNA*, September 2002.

[16] N. Bulusu, J. Heidemann, and D. Estrin, "GPS-less Low Cost Outdoor Localization for Very Small Devices," *IEEE Personal Communications Magazine*, 7, 5, October 2000, 28–34.

[17] CC1000 Data Sheet, online at www.chipcon.com/files/CC1000_Data_Sheet_2_2.pdf

[18] CC2420 Data Sheet, online at www.chipcon.com/files/CC2420_Data_Sheet_1_2.pdf

[19] A. Cerpa and D. Estrin, "ASCENT: Adaptive Self-Configuring Sensor Networks Topologies," *Proceedings of IEEE INFOCOM*, June 2002.

[20] J.-H. Chang and L. Tassiulas, "Energy Conserving Routing in Wireless Ad-hoc Networks," *Proceedings of IEEE INFOCOM*, March 2000.

[21] J.-H. Chang and L. Tassiulas, "Maximum Lifetime Routing in Wireless Sensor Networks," *IEEE/ACM Transactions on Networking*, 12, 4, August 2004, 609–619.

[22] N. Chang and M. Liu, "Revisiting the TTL-Based Controlled Flooding Search: Optimality and Randomization," *Proceedings of ACM MobiCom*, September 2004.

[23] B. Chen, K. Jamieson, H. Balakrishnan, and R. Morris, "Span: An Energy-Efficient Coordination Algorithm for Topology Maintenance in Ad Hoc Wireless Networks," *Proceedings of ACM MobiCom*, July 2001.

[24] Z. Cheng and W. Heinzelman, "Flooding Strategy for Target Discovery in Wireless Networks," *Proceedings of the Sixth ACM International Workshop on Modeling, Analysis and Simulation of Wireless and Mobile Systems (MSWiM)*, September 2003.

[25] E. Cheong, J. Liebman, J. Liu, and F. Zhao, "TinyGALS: A Programming Model for Event-Driven Embedded Systems," *Proceedings of the 18th ACM Symposium on Applied Computing (SAC'03)*, March 2003.

[26] K. Chintalapudi and R. Govindan, "Localized Edge Detection in a Sensor Field," *Proceedings of IEEE SNPA*, May 2003.

[27] K. Chintalapudi, A. Dhariwal, R. Govindan, and G. Sukhatme, "Localization Using Ranging and Sectoring," *Proceedings of IEEE INFOCOM*, March 2004.

[28] M. Chu, H. Haussecker, and F. Zhao, "Scalable Information-Driven Sensor Querying and Routing for Ad Hoc Heterogeneous Sensor Networks," *International Journal of High Performance Computing Applications*, 16, 3, August 2002, 293–313.

[29] I. Cidon and M. Sidi, "Distributed Assignment Algorithms for Multi-Hop Packet-Radio Networks," *IEEE Transactions on Computers*, 38, 10, Ocotber 1989, 1353–1361.

[30] T. Clouqueur, V. Phipatanasuphorn, P. Ramanathan, and K. K. Saluja, "Sensor Deployment Strategy for Target Detection," *Proceedings of ACM WSNA*, September 2002.

[31] S. Coleri, A. Puri, and P. Varaiya, "Power Efficient system for Sensor Networks," *Proceedings of Eighth IEEE International Symposium on Computers and Communication (ISCC)*, July 2003.

[32] W. S. Conner, J. Chhabra, M. Yarvis, and L. Krishnamurthy, "Experimental Evaluation of Synchronization and Topology Control for In-Building Sensor Network Applications," *Proceedings of ACM WSNA*, September 2003.

[33] P. I. Corke, S. E. Hrabar, R. Peterson, D. Rus, S. Saripalli, and G. S. Sukhatme, "Deployment and Connectivity Repair of a Sensor Net," *Proceedings of the 9th International Symposium on Experimental Robotics (ISER)*, June 2004.

[34] T. M. Cover and J. A. Thomas, *Elements of Information Theory*, John Wiley, 1991.

[35] R. Cristescu, B. Beferull-Lozano, and M. Vetterli, "On Network Correlated Data-Gathering," *Proceedings of IEEE INFOCOM*, March 2004.

[36] F. Cristian, "A Probabilistic Approach to Distributed Clock Synchronization," *Distributed Computing*, 3, 1989, 146–158.

[37] T. V. Dam and K. Langendoen, "An Adaptive Energy-Efficient MAC Protocol for Wireless Sensor Networks," *Proceedings of ACM SenSys*, November 2003.

[38] A. D'Costa and A. M. Sayeed, "Collaborative Signal Processing for Distributed Classification in Sensor Networks," *Proceedings of IPSN*, April 2003.

[39] D. S. J. De Couto, D. Aguayo, J. Bicket, and R. Morris, "A High-Throughput Path Metric for Multi-Hop Wireless Routing," *Proceedings of ACM MobiCom*, September 2003

[40] L. Doherty, K. S. J. Pister, and L. E. Ghaoui, "Convex Position Estimation in Wireless Sensor Networks," *Proceedings of IEEE INFOCOM*, April 2001.

[41] R. M. D'Souza, D. Galvin, C. Moore, and D. Randall, "Building Up a Connected Sparse Graph Using Local Geometric Information," personal communication 2004.

[42] C.-T. Ee and R. Bajcsy, "Congestion Control and Fairness for Many-to-One Routing in Sensor Networks," *Proceedings of ACM SenSys*, November 2004.

[43] J. Elson, L. Girod, and D. Estrin, "Fine-Grained Network Time Synchronization using Reference Broadcasts," in *Proceedings of the Fifth Symposium on Operating Systems Design and Implementation (OSDI)*, December 2002.

[44] L. Girod, J. Elson, A. Cerpa, T. Stathopoulos, N. Ramanathan, and D. Estrin, "EmStar: a Software Environment for Developing and Deploying Wireless Sensor Networks," *Proceedings of USENIX General Track*, June–July 2004.

[45] M. Enachescu, A. Goel, R. Govindan, and R. Motwani, "Scale Free Aggregation in Sensor Networks," *Proceedings of ALGOSENSORS*, July 2004.

[46] T. Eren, D. Goldenberg, W. Whiteley, Y. R. Yang, A. S. Morse, B. D. O. Anderson, and P. N. Belhumeur, "Rigidity, Computation, and Randomization in Network Localization," *Proceedings of IEEE INFOCOM*, March 2004.

[47] D. Estrin et al., "Embedded, Everywhere: A Research Agenda for Networked Systems of Embedded Computers," A National Research Council Report, National Academy Press, 2001.

[48] Q. Fang, F. Zhao, and L. Guibas, "Lightweight Sensing and Communication Protocols for Target Enumeration and Aggregation," *Proceedings of ACM MobiHoc*, June 2003.

[49] Q. Fang, J. Gao, and L. Guibas, "Locating and Bypassing Routing Holes in Sensor Networks," *Proceedings of IEEE INFOCOM*, March 2004.

[50] A. Faradjian, J. E. Gehrke, and P. Bonnet, "GADT: A Probability Space ADT For Representing and Querying the Physical World," *Proceedings of the 18th International Conference on Data Engineering (ICDE)*, February 2002.

[51] J. Faruque and A. Helmy, "RUGGED: RoUting on finGerprint Gradients in sEnsor Networks," *Proceedings of the IEEE International Conference on Pervasive Services (ICPS)*, July 2004.

[52] K. Finkenzeller, *RFID Handbook: Fundamentals and Applications in Contactless Smart Cards and Identification*, John Wiley & Sons, 2003.

[53] G. Finn, "Routing and Addressing Problems in Large Metropolitan-Scale Internetworks," ISI Research Report ISU/RR-87–180, March 1987.

[54] A. Galstyan, B. Krishnamachari, K. Lerman, and S. Pattem, "Distributed Online Localization in Sensor Networks Using a Moving Target," *Proceedings of International Symposium on Information Processing in Sensor Networks (IPSN)*, April 2004.

[55] S. Ganeriwal, R. Kumar, and M. B. Srivastava, "Timing-Sync Protocol for Sensor Networks," *Proceedings of ACM SenSys'03*, November 2003.

[56] S. Ganeriwal, D. Ganesan, M. Hansen, M. B. Srivastava, and D. Estrin, "Predictive Time Synchronization for Long-Lived Sensor Networks," University of Massachusetts, Amherst, Technical Report TR-04-97, 2004.

[57] D. Ganesan, B. Krishnamachari, A. Woo, D. Culler, D. Estrin, and S. Wicker, "Complex Behavior at Scale: An Experimental Study of Low-Power Wireless Sensor Networks," UCLA CS Technical Report UCLA/CSD-TR 02-0013, 2002.

[58] D. Ganesan, R. Govindan, S. Shenker, and D. Estrin, "Highly-Resilient, Energy-Efficient Multipath Routing in Wireless Sensor Networks," *Mobile Computing and Communications Review (MC2R)*, 1, 2, 2002.

[59] D. Ganesan, D. Estrin, and J. Heidemann, "DIMENSIONS: Why Do We Need a New Data Handling Architecture for Sensor Networks?" *Proceedings of the ACM Workshop on Hot Topics in Networks (HotNets-I)*, October 2002.

[60] D. Ganesan, B. Greenstein, D. Perelyubskiy, D. Estrin, and J. Heidemann, "An Evaluation of Multi-resolution Storage for Sensor Networks," *Proceedings of ACM SenSys*, November 2003.

[61] L. Girod and D. Estrin, "Robust Range Estimation using Acoustic and Multimodal Sensing," *Proceedings of IEEE/RSJ IROS*, October–November 2001.

[62] O. Gnawali, M. Yarvis, J. Heidemann, and R. Govindan, "Interaction of Retransmission, Blacklisting, and Routing Metrics for Reliability in Sensor Network Routing," *Proceedings of IEEE SECON*, October 2004.

[63] S. Goel and T. Imielinski, "Prediction-Based Monitoring in Sensor Networks: Taking Lessons from MPEG," *Computer Communication Review*, 31, 5, 2001, 82–95.

[64] A. Goel and D. Estrin, "Simultaneous Optimization for Concave Costs: Single Sink Aggregation or Single Source Buy-at-bulk," *Proceedings of the ACM-SIAM Symposium on Discrete Algorithms (SODA)*, January 2003.

[65] A. Goel, S. Rai, and B. Krishnamachari, "Sharp Thresholds for Monotone Properties in Random Geometric Graphs," *Proceedings of ACM Symposium on Theory of Computing (STOC)*, June 2004.

[66] R. Govindan, J. M. Hellerstein, W. Hong, S. Madden, M. Franklin, and S. Shenker, "The Sensor Network as a Database," USC Technical Report No. 02-771, September 2002.

[67] B. Greenstein, D. Estrin, R. Govindan, S. Ratnasamy, and S. Shenker, "DIFS: A Distributed Index for Features in Sensor Networks," *Proceedings of the First IEEE International Workshop on Sensor Network Protocols and Applications (SNPA)*, May 2003.

[68] J. V. Greunen and J. Rabaey, "Lightweight Time Synchronization for Sensor Networks," *Proceedings of ACM WSNA*, September 2003.

[69] C. Guestrin, P. Bodik, R. Thibaux, M. Paskin, and S. Madden, "Distributed Regression: an Efficient Framework for Modeling Sensor Network Data," *Proceedings of IPSN*, April 2004.

[70] P. Gupta and P. R. Kumar, "The Capacity of Wireless Networks," *IEEE Transactions on Information Theory*, 46, 2, March 2000, 388–404.

[71] P. Gupta and P. R. Kumar, "Critical Power for Asymptotic Connectivity in Wireless Networks," in W. M. McEneany *et al.* (eds), *Stochastic Analysis, Control, Optimization and Applications*, Birkhauser, 1998, pp. 547–566.

[72] T. He, C. Huang, B. M. Blum, J. A. Stankovic, and T. Abdelzaher, "Range-Free Localization Schemes for Large Scale Sensor Networks," *ACM MobiCom*, September 2003.

[73] T. He, J. A. Stankovic, C. Lu, and T. F. Abdelzaher, "SPEED: A Stateless Protocol for Real-Time Communication in Sensor Networks," *Proceedings of IEEE ICDCS*, May 2003

[74] J. Heidemann, F. Silva, C. Intanagonwiwat, R. Govindan, D. Estrin, and D. Ganesan, "Building Efficient Wireless Sensor Networks with Low-Level Naming," *Proceedings of the ACM Symposium on Operating Systems Principles (SOSP)*, October 2001.

[75] J. Heidemann, F. Silva, and D. Estrin, "Matching Data Dissemination Algorithms to Application Requirements," *Proceedings of ACM SenSys*, November 2003.

[76] W. R. Heinzelman, A. Chandrakasan, and H. Balakrishnan, "Energy-Efficient Communication Protocol for Wireless Microsensor Networks," *Proceedings of the Hawaii International Conference on System Sciences (HICSS)*, January 2000.

[77] W. B. Heinzelman, A. L. Murphy, H. S. Carvalho, and M. A. Perillo, "Middleware to Support Sensor Network Applications," *IEEE Network Magazine, Special Issue on Middleware Technologies for Future Communication Networks* 18, 1, Janurary 2004, 6–14.

[78] J. M. Hellerstein, W. Hong, S. Madden, and K. Stanek, "Beyond Average: Towards Sophisticated Sensing with Queries," *Proceedings of IPSN*, March 2003.

[79] N. Heo and P. K. Varshney, "An Intelligent Deployment and Clustering Algorithm for a Distributed Mobile Sensor Network," *Proceedings of IEEE International Conference on Systems, Man and Cybernetics*, vol. 5, October 2003.

[80] J. Hightower and G. Borriello, "Location Systems for Ubiquitous Computing," *IEEE Computer*, 34, 8, August 2001, 57–66.

[81] J. Hill, R. Szewcyk, A. Woo, D. Culler, S. Hollar, and K. Pister, "System Architecture Directions for Networked Sensors," *Proceedings of the 9th International Conference on Architectural Support for Programming Languages and Operating Systems (ASPLOS)*, November 2000.

[82] J. Hill and D. Culler, "Mica: A Wireless Platform for Deeply Embedded Networks," *IEEE Micro*, 22, 6, November–December 2002, 12–24.

[83] A. El-Hoiydi, "Aloha with Preamble Sampling for Sporadic Traffic in Ad Hoc Wireless Sensor Networks," *Proceedings of IEEE International Conference on Communications (ICC)*, April 2002.

[84] A. El-Hoiydi and J. D. Decotignie, "WiseMAC: An Ultra Low-Power MAC Protocol for Multi-Hop Wireless Sensor Networks," *Proceedings of First International Workshop on Algorithmic Aspects of Wireless Sensor Networks (ALGOSENSORS)*, July 2004.

[85] T. C. Hou and V. O. K. Li, "Transmission Range Control in Multi-hop Packet Radio Networks," *IEEE Transactions on Communication*, 34, 1, January 1986, 38–44.

[86] A. Howard, M. Mataric, and G. Sukhatme, "Relaxation on a Mesh: A Formalism for Generalized Localization," *Proceedings of IEEE/RSJ IROS*, October–November 2001.

[87] A. Howard, M. J. Mataric, and G. S. Sukhatme, "Mobile Sensor Network Deployment Using Potential Fields: A Distributed, Scalable Solution to the Area Coverage Problem," in H. Asama and T. Arai and T. Fukuda and T. Hasegawa (eds), *Distributed Autonomous Robotic Systems*, Springer, 2002, 299–308.

[88] A. Howard, M. J. Mataric, and G. S. Sukhatme, "An Incremental Deployment Algorithm for Mobile Robot Teams," *Proceedings of IEEE/RSJ International Conference on Robotics and Intelligent Systems (IROS)*, October 2002.

[89] A. Hu and S. D. Servetto, "Asymptotically Optimal Time Synchronization in Dense Sensor Networks," *Proceedings of ACM WSNA*, September 2003.

[90] W. Hu, C.-T. Chou, S. Jha, and N. Bulusu, "Deploying Long-Lived and Cost-Effective Hybrid Sensor Networks," *Proceedings of the First Workshop on Broadband Advanced Sensor Networks (BaseNets)*, October 2004.

[91] Z. Hu and B. Li, "Fundamental Performance Limits of Wireless Sensor Networks," in Yang Xiao and Yi Pan (eds), *Ad Hoc and Sensor Networks*, Nova Science Publishers, 2004.

[92] C. Huang and Y. Tseng, "The Coverage Problem in a Wireless Sensor Network," *Proceedings of ACM WSNA*, September 2003.

[93] B. Hull, K. Jamieson, and H. Balakrishnan, "Techniques for Mitigating Congestion in Sensor Networks," *Proceedings of ACM SenSys*, November 2004.

[94] *IEEE 802.15.4, IEEE Standard for Information Technology-Part 15.4: Wireless Medium Access Control (MAC) and Physical Layer (PHY) Specifications for Low Rate Wireless Personal Area Networks (LR-WPANS)*, 2003.

[95] *IEEE 802.11, IEEE Standards for Information Technology – Telecommunications and Information Exchange between Systems – Local and Metropolitan Area Network – Specific Requirements – Part 11: Wireless LAN Medium Access Control (MAC) and Physical Layer (PHY) Specifications*, 1999.

[96] C. Intanagonwiwat, R. Govindan, and D. Estrin, "Directed Diffusion: A Scalable and Robust Communication Paradigm for Sensor Networks," *Proceedings of ACM MobiCom*, August 2000.

[97] C. Intanagonwiwat, R. Govindan, D. Estrin, J. Heidemann, and F. Silva, "Directed Diffusion for Wireless Sensor Networking," *ACM/IEEE Transactions on Networking*, 11, 1, February 2002.

[98] W. J. Kaiser, K. Bult, A. Burstein, D. Chang, *et al.*, "Wireless Integrated Microsensors," *Technical Digest of the 1996 Solid State Sensor and Actuator Workshop*, June 1996.

[99] K. Kalpakis, K. Dasgupta, and P. Namjoshi, "Efficient Algorithms for Maximum Lifetime Data-Gathering and Aggregation in Wireless Sensor Networks," *Computer Networks: The International Journal of Computer and Telecommunications Networking*, 42, 6, August 2003, 697–716.

[100] C. Karlof and D. Wagner, "Secure Routing in Wireless Sensor Networks: Attacks and Countermeasures," *Proceedings of IEEE SNPA*, May 2003.

[101] P. Karn, "MACA: A New Channel Access Method for Packet Radio," *Proceedings of the 9th ARRL/CRRL Amateur Radio Computer Networking Conference*, September 1990.

[102] Brad Karp and H. T. Kung, "GPSR: Greedy Perimeter Stateless Routing for Wireless Networks," *Proceedings of ACM MobiCom*, August 2000.

[103] R. Karp, J. Elson, D. Estrin, and S. Shenker, "Optimal and Global Time Synchronization in Sensornets," CENS Technical Report 0012, April 2003.

[104] V. Kawadia and P. R. Kumar, "Power Control and Clustering in Ad Hoc Networks," *Proceedings of IEEE INFOCOM*, April 2003.

[105] E. Khandani, J. Abounadi, E. Modiano, and L. Zheng, "Cooperative Routing in Wireless Networks," *Proceedings of Allerton Conference*, March 2004.

[106] E. Khandani, E. Modiano, J. Abounadi, and L. Zheng, "Reliability and Route Diversity in Wireless Networks," MIT LIDS Technical Report Number 2634, November 2004.

[107] H. S. Kim, T. F. Abdelzaher, and W. H. Kwon, "Dissemination: Minimum Energy Asynchronous Dissemination to Mobile Sinks in Wireless Sensor Networks," *Proceedings of ACM SenSys*, November 2003.

[108] H. Kopetz and W. Schwabl, "Global Time in Distributed Real-Time Systems," Technical Report 15/89, Technische Universitat Wien, Wien Austria, October 1989.

[109] B. Krishnamachari, D. Estrin, and S. B. Wicker, "The Impact of Data Aggregation in Wireless Sensor Networks," *Proceedings of the ICDCS Workshop on Distributed Event-based Systems (DEBS)*, July 2002.

[110] B. Krishnamachari and F. Ordonez, "Fundamental Limits of Networked Sensing," in C. S. Raghavendra (eds), K. Sivalingam, and T. Znati *Wireless Sensor Networks*, Kluwer Academic Publishers, 2004, 233–249.

[111] B. Krishnamachari and J. Heidemann, "Application-specific Modelling of Information Routing in Sensor Networks," *Proceedings of the IEEE IPCCC Workshop on Multi-hop Wireless Networks (MWN)*, April 2004.

[112] L. Lamport, "Time, Clocks, and the Ordering of Events in a Distributed System," *Communications of the ACM*, 21, 7, July 1978, 558–565.

[113] K. Langendoen and N. Reijers, "Distributed Localization in Wireless Sensor Networks: A Quantitative Comparison," Parallel and Distributed Systems Report Series, Technical Report Number PDS-2002-3, Delft University of Technology, 2002.

[114] J.-J. Lee, B. Krishnamachari and C.-C. J. Kuo, "Impact of Heterogeneous Deployment on Lifetime Sensing Coverage in Sensor Networks," First IEEE International Conference on Sensor and Ad hoc Communications and Networks (SECON), October 2004.

[115] P. Levis and D. Culler, "Mate: A Tiny Virtual Machine for Sensor Networks," *Proceedings of ASPLOS*, October 2002.

[116] L. Li and J. Y. Halpern, "Minimum Energy Mobile Wireless Networks Revisited," *Proceedings of the IEEE International Conference on Communications (ICC)*, June 2001.

[117] L. Li, J. Y. Halpern, V. Bahl, Y.-M. Wang, and R. Wattenhofer, "Analysis of a Cone-Based Distributed Topology Control Algorithms for Wireless Multi-hop Networks," *Proceedings of ACM Symposium on Principles of Distributed Computing (PODC)*, August 2001.

[118] N. Li, J. Hou, and L. Sha, "Design and Analysis of an MST-Based Topology Control Algorithm," *Proceedings of IEEE INFOCOM*, April 2003.

[119] X. Li, Y. J. Kim, R. Govindan, and W. Hong, "Multi-dimensional Range Queries in Sensor Networks," *Proceedings of ACM Sensys*, November 2003.

[120] E. A. Lin, J. M. Rabaey, and A. Wolisz, "Power-Efficient Rendezvous Schemes for Dense Wireless Sensor Networks," *Proceedings of ICC*, June 2004.

[121] J. Liu, P. Cheung, L. Guibas, and F. Zhao, "A Dual-Space Approach to Tracking and Sensor Management in Wireless Sensor Networks," *Proceedings of ACM WSNA*, September 2002.

[122] J. Liu, F. Zhao, and D. Petrovic, "Information-Directed Routing in Ad Hoc Sensor Networks," *Proceedings of ACM WSNA*, September 2003.

[123] J. Liu, M. Chu, J. Liu, J. Reich, and F. Zhao, "State-Centric Programming for Sensor-Actuator Network Systems," *IEEE Pervasive Computing*, 2, 4, October–December 2003, 50–62.

[124] X. Liu, Q. Huang, and Y. Zhang, "Combs, Needles, Haystacks: Balancing Push and Pull for Discovery in Large-Scale Sensor Networks," *Proceedings of ACM Sensys*, November 2004.

[125] C. Lu, B. M. Blum, T. F. Abdelzaher, J. A. Stankovic, and T. He, "RAP: A Real-Time Communication Architecture for Large-Scale Wireless Sensor Networks," *Proceedings of IEEE Real-Time and Embedded Technology and Applications Symposium (RTAS)*, September 2002.

[126] G. Lu, B. Krishnamachari, and C. Raghavendra, "Performance Evaluation of the IEEE 802.15.4 MAC for Low-Rate Low-Power Wireless Networks," *Proceedings of the IEEE IPCCC Workshop on Energy-Efficient Wireless Communications and Networks (EWCN)*, April 2004.

[127] G. Lu, B. Krishnamachari, and C. Raghavendra, "An Adaptive Energy-Efficient and Low-Latency MAC for Data-Gathering in Sensor Networks," *Proceedings of the 4th International IEEE Workshop on Algorithms for Wireless, Mobile, Ad Hoc and Sensor Networks (WMAN)*, April 2004.

[128] G. Lu, N. Sadagopan, B. Krishnamachari, and A. Goel, "Delay Efficient Sleep Scheduling in Wireless Sensor Networks," *Proceedings of IEEE INFOCOM*, March 2005.

[129] S. Madden, M. J. Franklin, J. M. Hellerstein, and W. Hong, "TAG: A Tiny AGgregation Service for AdHoc Sensor Networks," *Proceedings of the 5th Symposium on Operating System Design and Implementation (OSDI)*, December 2002.

[130] A. Mainwaring, J. Polastre, R. Szewczyk, D. Culler, and J. Anderson, "Wireless Sensor Networks for Habitat Monitoring," *Proceedings of the First ACM International Workshop on Wireless Sensor Networks and Applications (WSNA)*, Atlanta, Georgia, September 2002.

[131] W. Manges, "It's Time for Sensors to Go Wireless," *Sensors Magazine*, April 1999.

[132] A. Manjeshwar and D. P. Agrawal, "TEEN: A Routing Protocol for Enhanced Efficiency in Wireless Sensor Networks," *Proceedings of the 15th International Parallel and Distributed Processing Symposium (IPDPS)*, April 2001.

[133] D. Marco, E. J. Duarte-Melo, M. Liu, and D. L. Neuhoff, "On the Many-to-One Transport Capacity of a Dense Wireless Sensor Network and the Compressibility of Its Data," *Proceedings of IPSN*, April 2003.

[134] M. Maroti, B. Kusky, G. Simon, and A. Ledeczi, "The Flooding Time Synchronization Protocol," *Proceedings of ACM SenSys*, November 2004.

[135] M. Mauve, J. Widmer, and H. Hartenstein, "A Survey on Position-Based Routing in Mobile Ad Hoc Networks," *IEEE Network Magazine*, 15, 6, November 2001, 30–39.

[136] S. Meguerdichian, F. Koushanfar, M. Potkonjak, and M. B. Srivastava, "Coverage Problems in Wireless Ad Hoc Sensor Networks," *Proceedings of IEEE INFOCOM*, April 2001.

[137] S. Meguerdichian, F. Koushanfar, G. Qu, and M. Potkonjak, "Exposure in Wireless Ad Hoc Sensor Networks," *Proceedings of ACM MobiCom*, July 2001.

[138] D. L. Mills, "Internet Time Synchronization: The Network Time Protocol," *IEEE Transactions on Communications*, 39, 10, October 1991, 1482–1493.

[139] R. Min and A. Chandrakasan, "A Framework for Energy-Scalable Communication in High-Density Wireless Networks," *Proceedings of the International Symposium on Low-Power Electronics and Design (ISLPED)*, August 2002.

[140] R. Min and A. Chandrakasan, "Top Five Myths about the Energy Consumption of Wireless Communication," *ACM Mobile Computing and Communications Review*, 7, 1, January 2003, 65–67.

[141] P. Misra and P. Enge, *Global Positioning System: Signals, Measurements, and Performance*, Ganga-Jamuna Press, 2001.

[142] S. Narayanaswamy, V. Kawadia, R. S. Sreenivas, and P. R. Kumar, "Power Control in Ad-Hoc Networks: Theory, Architecture, Algorithm and Implementation of the COMPOW Protocol," *Proceedings of the European Wireless Conference – Next Generation Wireless Networks: Technologies, Protocols, Services and Applications*, February 2002.

[143] A. Nasipuri and K. Li, "A Directionality-Based Location Discovery Scheme for Wireless Sensor Networks," *Proceedings of ACM WSNA*, September 2002.

[144] S. Nath, P. B. Gibbons, S. Seshan, and Z. Anderson, "Synopsis Diffusion for Robust Aggregation in Sensor Networks," *Proceedings of ACM SenSys*, November 2004.

[145] D. Gay, P. Levis, R. von Behren, M. Welsh, E. Brewer, and D. Culler, "The nesC Language: A Holistic Approach to Networked Embedded Systems," *Proceedings of Programming Language Design and Implementation (PLDI)*, June 2003.

[146] D. Niculescu and B. Nath, "Ad Hoc Positioning System (APS)," *Proceedings of the IEEE Global Communications Conference (GLOBECOM)*, November 2001.

[147] D. Niculescu and B. Nath, "Ad Hoc Positioning System (APS) Using AOA," *Proceedings of IEEE INFOCOM*, April 2003.

[148] D. Niculescu and B. Nath, "Trajectory-Based Forwarding and Its Applications," *Proceedings of ACM MobiCom*, September 2003.

[149] H. Nikookar and H. Hashemi, "Statistical Modeling of Signal Amplitude Fading of Indoor Radio Propagation Channels," *Proceedings of Second International Conference on Universal Personal Communications*, October 1993.

[150] R. Nowak and U. Mitra, "Boundary Estimation in Sensor Networks: Theory and Methods," *Proceedings of IPSN*, April 2003.

[151] J. Pan, Y. T. Hou, L. Cai, Y. Shi, and S. X. Shen, "Topology Control for Wireless Sensor Networks," *Proceedings of ACM MobiCom*, September 2003.

[152] S. Pattem, S. Poduri, and B. Krishnamachari, "Energy–Quality Tradeoffs for Target Tracking in Wireless Sensor Networks," *Proceedings of IPSN*, April 2003.

[153] S. Pattem, B. Krishnamachari, and R. Govindan, "The Impact of Spatial Correlation on Routing with Compression in Wireless Sensor Networks," *Proceedings of IPSN*, April 2004.

[154] N. Patwari, R. J. O'Dea, and Y. Wang, "Relative Location in Wireless Networks," *Proceedings of IEEE Vehicular Technology Conference (VTC-Spring)*, May 2001.

[155] N. Patwari and A. O. Hero III, "Using Proximity and Quantized RSS for Sensor Localization in Wireless Networks," *Proceedings of ACM WSNA*, September 2003.

[156] M. Penrose, "The Longest Edge of the Random Minimal Spanning Tree," *Annals of Applied Probability*, 7, 2, 1997, 340–361.

[157] M. Penrose, "On K-Connectivity for a Geometric Random Graph," *Random Structures and Algorithms*, 15, 2, 1999, 145–164.

[158] C. E. Perkins, *Ad Hoc Networking*, Addison-Wesley, 2001.

[159] A. Perrig, R. Szewczyk, V. Wen, D. E. Culler, and J. D. Tygar, "SPINS: Security Protocols for Sensor Networks," *Proceedings of ACM MobiCom*, 2001.

[160] S. Poduri and G. S. Sukhatme, "Constrained Coverage for Mobile Sensor Networks," *Proceedings of IEEE International Conference on Robotics and Automation (ICRA)*, May 2004.

[161] J. Polastre, J. Hill, and D. Culler, "Versatile Low-Power Media Access for Wireless Sensor Networks," *Proceedings of ACM Sensys*, November 2004.

[162] L. Pond and V. O. K. Li, "A Distributed Time-Slot Assignment Protocol for Mobile Multi-Hop Broadcast Packet Radio Networks," *Proceedings of IEEE Military Communications Conference (MILCOM)*, October 1989.

[163] R. Poor, C. Bowman, and C. B. Auburn, "Self-Healing Networks," *ACM Queue*, 1, 3, May 2003, 52–59.

[164] N. B. Priyantha, A. Chakraborty, and H. Balakrishnan, "The Cricket Location-Support System," *Proceedings of ACM MobiCom*, August 2000.

[165] W. Quadeer, T. Simunic, J. Ankcorn, V. Krishnan, and G. De Micheli, "Heterogeneous Wireless Network Management," HP Labs Technical Report HPL-2003–252, December 2003.

[166] J. Rabaey, M. Josie Ammer, J. L. da Silva, D. Patel and S. Roundy, "PicoRadio Supports Ad Hoc Ultra-Low-Power Wireless Networking," *IEEE Computer Magazine*, 33, 7, July 2000, 42–48.

[167] V. Rajendran, K. Obraczka, and J. J. Garcia-Luna-Aceves, "Energy-Efficient, Collision-Free Medium Access Control for Wireless Sensor Networks," *Proceedings of ACM Sensys*, November 2003.

[168] S. Ramanathan and R. Rosales-Hain, "Topology Control of Multi-hop Radio Networks Using Transmit Power Adjustment," *Proceedings of IEEE INFOCOM*, March 2000.

[169] R. Ramaswami and K. K. Parhi, "Distributed Scheduling of Broadcasts in a Radio Network," *Proceedings of IEEE INFOCOM*, April 1989.

[170] A. Rao, S. Ratnasamy, C. Papadimitriou, S. Shenker, and I. Stoica, "Geographic Routing without Location Information," *Proceedings of ACM MobiCom*, September 2003.

[171] T. S. Rappaport, *Wireless Communications: Principles and Practice*, Prentice Hall, 2002.

[172] S. Ratnasamy, B. Karp, S. Shenker, D. Estrin, R. Govindan, L. Yin, and F. Yu, "Data-Centric Storage in Sensornets with GHT: A Geographic Hash Table," *Mobile Networks and Applications (MONET), Special Issue on Wireless Sensor Networks*, 8, 4, August 2003, 427–442.

[173] S. Ray, R. Ungrangsi, F. D. Pellegrini, A. Trachtenberg, and D. Starobinski, "Robust Location Detection in Emergency Sensor Networks," *IEEE INFOCOM*, April 2003.

[174] V. Rodoplu and T. H. Meng, "Minimum Energy Mobile Wireless Networks," *IEEE Journal on Selected Areas in Communications (JSAC)*, 17, 8, August 1999, 1333–1344.

[175] K. Romer, O. Kasten, and F. Mattern, "Middleware Challenges for Wireless Sensor Networks," *ACM Mobile Computing and Communication Review*, 6, 4, October 2002, 59–61.

[176] L. Rossi, B. Krishnamachari, and C. C. Jay Kuo, "Distributed Parameter Estimation for Monitoring Diffusion Phenomena Using Physical Models," *Proceedings of IEEE SECON*, October 2004.

[177] N. Sadagopan, B. Krishnamachari, and A. Helmy, "The ACQUIRE Mechanism for Efficient Querying in Sensor Networks," *Proceedings of the First IEEE International Workshop on Sensor Network Protocols and Applications (SNPA)*, May 2003.

[178] N. Sadagopan, B. Krishnamachari, and A. Helmy, "Active Query Forwarding in Sensor Networks (ACQUIRE)," *Elsevier Journal of Ad Hoc Networks*, 2004.

[179] N. Sadagopan and B. Krishnamachari, "Maximizing Data Extraction in Energy-Limited Sensor Networks," *Proceedings of IEEE INFOCOM*, March 2004.

[180] Y. Sankarasubramaniam, I. F. Akyildiz, and S. W. McLaughlin, "Energy Efficiency based Packet Size Optimization in Wireless Sensor Networks," *Proceedings of the First IEEE International Workshop on Sensor Network Protocols and Applications (SNPA)*, May 2003.

[181] Y. Sankarasubramaniam, O. B. Akan, and I. F. Akyildiz, "ESRT: Event-to-Sink Reliable Transport in Wireless Sensor Networks," *Proceedings of ACM MobiHoc*, June 2003.

[182] C. Savarese, J. Rabaey, and K. Langendoen, "Robust Positioning Algorithms for Distributed Ad-Hoc Wireless Sensor Networks," *USENIX Annual Technical Conference*, 2001.

[183] A. Savvides, C. C. Han, and M. B. Srivastava, "Dynamic Fine Grained Localization in Ad-Hoc Sensor Networks," *Proceedings of ACM MobiCom*, July 2001.

[184] A. Savvides, W. Garber, S. Adlakha, R. Moses, and M. B. Srivastava, "On the Error Characteristics of Multi-hop Node Localization in Ad-Hoc Sensor Networks," *Proceedings of IPSN*, April 2003.

[185] A. Savvides, M. Srivastava, L. Girod, and D. Estrin, "Localization in Sensor Networks," C. S. Raghavendra, K. Sivalingam and T. Znati (eds), in *Wireless Sensor Networks*, Kluwer, 2004.

[186] A. Scaglione and S. D. Servetto, "On the Interdependence of Routing and Data Compression in Multi-Hop Sensor Networks," *Proceedings of ACM MobiCom*, September 2002.

[187] K. Seada, M. Zuniga, A. Helmy, and B. Krishnamachari, "Energy-Efficient Forwarding Strategies for Geographic Routing in Wireless Sensor Networks," *Proceedings of ACM SenSys*, November 2004.

[188] DARPA IPTO, "Sensit: Sensor Information Technology Program," online at http://www.darpa.mil/ipto/programs/sensit/

[189] R. C. Shah and J. Rabaey, "Energy Aware Routing for Low Energy Ad Hoc Sensor Networks," *Proceedings of IEEE WCNC*, March 2002.

[190] R. C. Shah, S. Roy, S. Jain, and W. Brunette, "Data MULEs: Modeling a Three-Tier Architecture for Sparse Sensor Networks," *Proceedings of IEEE SNPA*, May 2003.

[191] S. Shakkottai, R. Srikant, and N. Shroff, "Unreliable Sensor Grids: Coverage, Connectivity and Diameter," *Proceedings of the 22nd Annual Joint Conference of the IEEE Computer and Communications Societies (INFOCOM)*, April 2003.

[192] S. Shakkottai, "Asymptotics of Query Strategies over a Sensor Network," *Proceedings of IEEE INFOCOM*, March 2004.

[193] Y. Shang, W. Rumi, and Y. Zhang, "Localization from Mere Connectivity," *Proceedings of ACM MobiHoc*, June 2003.

[194] M. A. Sharaf, J. Beaver, A. Labrinidis, and P. K. Chrysanthis, "TiNA: A Scheme for Temporal Coherency-Aware In-Network Aggregation," *Proceedings of the Third ACM International Workshop on Data Engineering for Wireless and Mobile Access (MobiDE)*, September 2003.

[195] E. Shih, P. Bahl, and M. J. Sinclair, "Wake On Wireless: An Event Driven Energy Saving Strategy for Battery Operated Devices," *Proceedings of ACM MobiCom*, September 2002.

[196] J. Shin, L. J. Guibas, and F. Zhao, "A Distributed Algorithm for Managing Multi-target Identities in Wireless Ad-Hoc Sensor Networks," *Proceedings of IPSN*, April 2003.

[197] N. Shrivastava, C. Buragohain, D. Agrawal, and S. Suri, "Medians and Beyond: New Aggregation Techniques for Sensor Networks," *Proceedings of ACM SenSys*, November 2004.

[198] M. L. Sichitiu and C. Veerarittiphan, "Simple, Accurate Time Synchronization for Wireless Sensor Networks," *Proceedings of IEEE Conference on Wireless Communications and Networking (WCNC)*, March 2003.

[199] S. Singh and C. S. Raghavendra, "PAMAS – Power Aware Multi-Access Protocol with Signalling for Ad Hoc Networks," *Proceedings of ACM MobiCom*, October 1998.

[200] S. Singh, M. Woo, and C. Raghavendra, "Power-Aware Routing in Mobile Ad Hoc Networks," *Proceedings of ACM MobiCom*, October 1998.

[201] S. Skiena, *Implementing Discrete Mathematics: Combinatorics and Graph Theory with Mathematica*, Addison-Wesley, 1990.

[202] S. Slijepcevic and M. Potkonjak, "Power Efficient Organization of Wireless Sensor Networks," *Proceedings of IEEE ICC*, June 2001.

[203] K. Sohrabi, J. Gao, V. Ailawadhi, and G. J. Pottie, "Protocols for Self-Organization of a Wireless Sensor Network," *IEEE Personal Communications*, 7, 5, October 2000.

[204] D. Son, B. Krishnamachari, and J. Heidemann, "Experimental Study of the Effects of Transmission Power Control and Blacklisting in Wireless Sensor Networks," *Proceedings of IEEE SECON*, October 2004.

[205] A. Sridharan and B. Krishnamachari, "Max–Min Fair Collision-Free Scheduling for Wireless Sensor Networks," *Proceedings of the IEEE IPCCC Workshop on Multi-hop Wireless Networks (MWN)*, April 2004.

[206] F. Stann and J. Heidemann, "RMST: Reliable Data Transport in Sensor Networks," *Proceedings of the First IEEE International Workshop on Sensor Network Protocols and Applications (SNPA)*, May 2003.

[207] G. L. Stuber, *Principles of Mobile Communication*, Kluwer, 1996.

[208] H. Takagi and L. Kleinrock, "Optimal Transmission Ranges for Randomly Distributed Packet Radio Terminals," *IEEE Transactions on Communications*, 32, 3, March 1984, 246–257.

[209] A. S. Tanenbaum and M. van Steen, *Distributed Systems: Principles and Paradigms*, Prentice Hall, 2002.

[210] D. Tian and N. D. Georganas, "A Coverage-Preserving Node Scheduling Scheme for Large Wireless Sensor Networks," *Proceedings of ACM WSNA*, September 2002.

[211] C. K. Toh, *Ad Hoc Mobile Wireless Networks*, Prentice Hall, 2002.

[212] P. Levis, N. Lee, M. Welsh, and D. Culler, "TOSSIM: Accurate and Scalable Simulation of Entier TinyOS Applications," *Proceedings of ACM Sensys*, November 2003.

[213] P. Verssimo, L. Rodrigues, and A. Casimiro, "CesiumSpray: A Precise and Accurate Global Time Service for Large-Scale Systems," *Kluwer Journal of Real-Time Systems, Special Issue on the Challenge of Global Time in Large-Scale Distributed Real-Time Systems*, 12, 3, November 1997.

[214] R. Virrankoski, "Localization in Ad-Hoc Sensor Networks," Unpublished report, Helsinki University of Technology, 2003.

[215] C.-Y. Wan, A. Campbell, and L. Krishnamurthy, "PSFQ: A Reliable Transport Protocol for Wireless Sensor Networks," *Proceedings of ACM WSNA*, September 2002.

[216] C.-Y. Wan, S. B. Eisenman, and A. T. Campbell, "CODA: Congestion Detection and Avoidance in Sensor Networks," *Proceedings of ACM SenSys*, November 2003.

[217] P.-J. Wan and C.-W. Yi, "Asymptotic Critical Transmission Radius and Critical Neighbor Number for K-Connectivity in Wireless Ad Hoc Networks," *Proceedings of the 5th ACM International Symposium on Mobile Ad Hoc Networking and Computing (MobiHoc)*, May 2004.

[218] S. S. Wang, M. Green, and M. Malkawa, "E-911 Location Standards and Location Commercial Services," *Proceedings of IEEE Emerging Technologies Symposium on Broadband, Wireless Internet Access*, April 2000.

[219] G. Wang, G. Cao, and T. La Porta, "A Bidding Protocol for Deploying Mobile Sensors," *Proceedings of 11th IEEE International Conference on Network Protocols (ICNP)*, November 2003.

[220] X. Wang, G. Xing, Y. Zhang, C. Lu, R. Pless, and C. Gill, "Integrated Coverage and Connectivity Configuration in Wireless Sensor Networks," *Proceedings of ACM SenSys*, November 2003.

[221] R. Want, A. Hopper, V. Falcao, and J. Gibbons, "The Active Badge Location System," *ACM Transactions on Information Systems*, 10, 1, January 1992, 91–102.

[222] R. Wattenhofer, L. Li, V. Bahl, and Y.-M. Wang, "Distributed Topology Control for Power Efficient Operation in Multi-hop Wireless Ad Hoc Networks," *Proceedings of IEEE INFOCOM*, April 2001.

[223] K. Whitehouse and D. Culler, "Calibration as Parameter Estimation in Sensor Networks," *Proceedings of ACM WSNA*, September 2002.

[224] A. Woo and D. Culler, "A Transmission Control Scheme for Media Access in Sensor Networks," *Proceedings of ACM Mobicom*, July 2001.

[225] A. Woo, T. Tong, and D. Culler, "Taming the Underlying Challenges of Reliable Multi-hop Routing in Sensor Networks," *Proceedings of ACM SenSys*, November, 2003.

[226] A. Woo, K. Whitehouse, F. Jiang, J. Polastre, and D. Culler, "The Shadowing Phenomenon: Implications of Receiving During a Collision," *UC Berkeley Technical Report UCB//CSD-04–1313*, March 2004.

[227] A. Wood and J. Stankovic, "Denial of Service in Sensor Networks," *IEEE Computer*, 35, 10, October 2002, 54–62.

[228] K. Wu, Y. Gao, F. Li, and Y. Xiao, "Lightweight Deployment-Aware Scheduling for Wireless Sensor Networks," *ACM/Kluwer MONET Journal*(to appear).

[229] Y. Xu, J. Heidemann, and D. Estrin, "Adaptive Energy-Conserving Routing for Multi-hop Ad Hoc Networks," *USC/InformationSciences Institute Technical Report 527*, October 2000.

[230] Y. Xu, J. Heidemann, and D. Estrin, "Geography-Informed Energy Conservation for Ad Hoc Routing," *Proceedings of ACM MobiCom*, July 2001.

[231] N. Xu, S. Rangwala, K. Chintalapudi, D. Ganesan, A. Broad, R. Govindan, and D. Estrin, "A Wireless Sensor Network for Structural Monitoring," *Proceedings of ACM Conference on Embedded Networked Sensor Systems (SenSys)*, November 2004.

[232] F. Xue and P. R. Kumar, "The Number of Neighbors Needed for Connectivity of Wireless Networks," *Wireless Networks*, 10, 2, March 2004, 169–181.

[233] Y. Yao and J. Gehrke, "The Cougar Approach to In-Network Query Processing in Sensor Networks," *SIGMOD Record*, 31, 1, March 2002, 9–18.

[234] W. Ye, J. Heidemann, and D. Estrin, "An Energy-Efficient MAC Protocol for Wireless Sensor Networks," *Proceedings of IEEE INFOCOM*, June 2002.

[235] F. Ye, G. Zhong, J. Cheng, S. Lu, and L. Zhang, "PEAS: A Robust Energy Conserving Protocol for Long-lived Sensor Networks," *Proceedings of IEEE International Conference on Distributed Computing Systems (ICDCS)*, 2003.

[236] F. Ye, G. Zhong, S. Lu, and L. Zhang, "A Robust Data Delivery Protocol for Large Scale Sensor Networks," *Proceedings of IPSN*, September 2003.

[237] W. Ye, J. Heidemann, and D. Estrin, "Medium Access Control with Coordinated, Adaptive Sleeping for Wireless Sensor Networks," *ACM/IEEE Transactions on Networking*, 12, 3, June 2004, 493–506.

[238] K. Yedavalli, B. Krishnamachari, S. Ravula, and B. Srinivasan, "Ecolocation: A Sequence-Based Technique for RF Localization in Wireless Sensor Networks," *Proceedings of IPSN*, April 2005.

[239] F. Ye, H. Luo, J. Cheng, S. Lu, and L. Zhang, "A Two-Tier Data Dissemination Model for Large-scale Wireless Sensor Networks," *Proceedings of ACM MobiCom*, September 2002.

[240] Y. Yu, R. Govindan, and D. Estrin, "Geographical and Energy Aware Routing: A Recursive Data Dissemination Protocol for Wireless Sensor Networks," *UCLA Computer Science Department Technical Report UCLA/CSD-TR-01–0023*, May 2001.

[241] Y. Yu, B. Krishnamachari, and V. K. Prasanna, "Issues in Designing Middleware for Wireless Sensor Networks," *IEEE Network Magazine, Special Issue on Middleware Technologies for Future Communication Networks*, 18, 1, January 2004, 15–21.

[242] J. Zhao and R. Govindan, "Understanding Packet Delivery Performance in Dense Wireless Sensor Networks," *Proceedings of ACM SenSys*, November 2003.

[243] R. Zheng, C. Hou, and L. Sha, "Asynchronous Wakeup for Ad Hoc Networks," *Proceedings of ACM MobiHoc*, June 2003.

[244] The Zigbee Alliance, online at http://www.zigbee.org

[245] M. Zuniga and B. Krishnamachari, "Analyzing the Transitional Region in Low-Power Wireless Links," *Proceedings of IEEE SECON*, October 2004.

Index